W9-CLJ-702

$ 5.65

Anthony

Structure and function of the body

Fourth edition

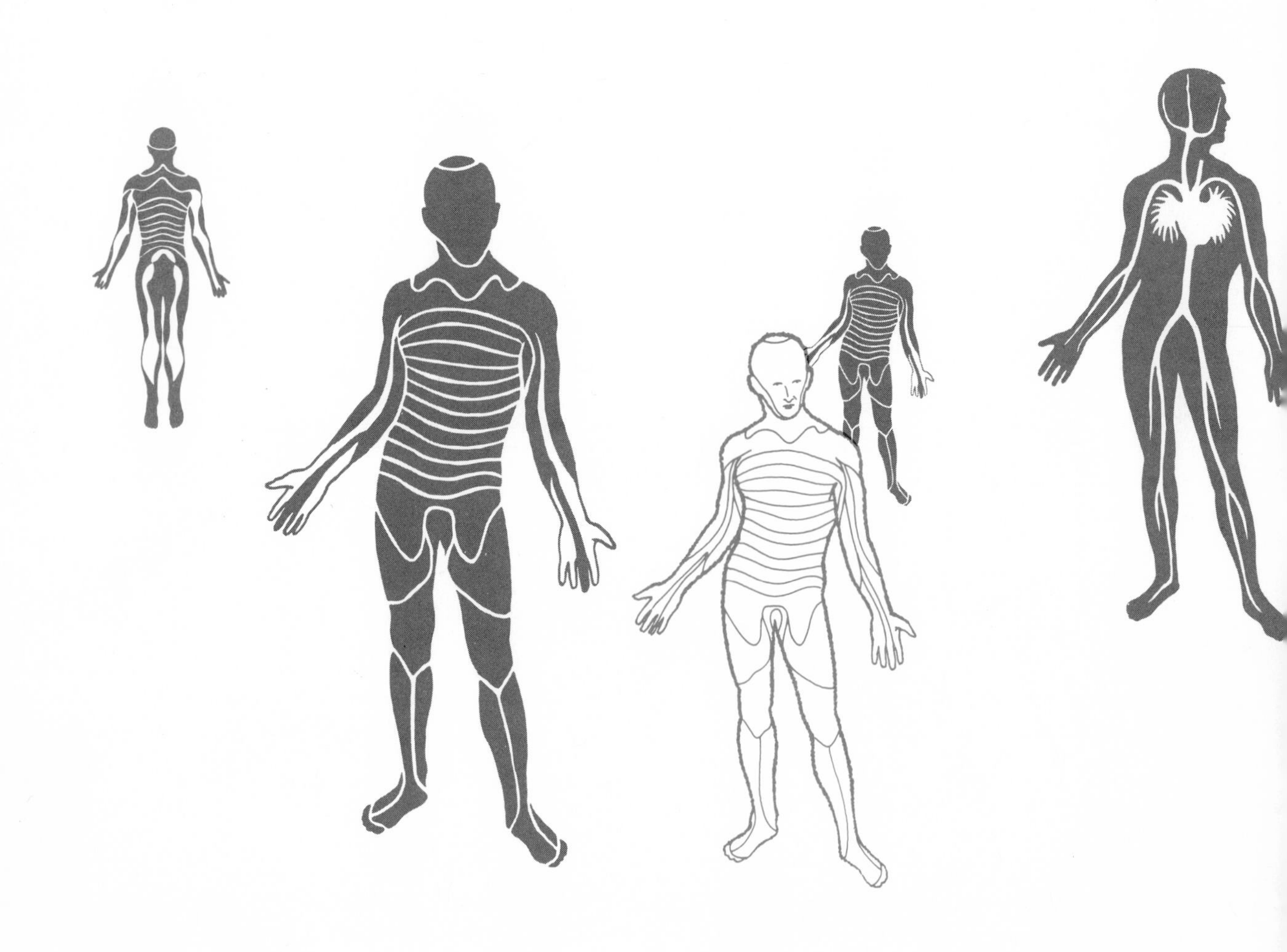

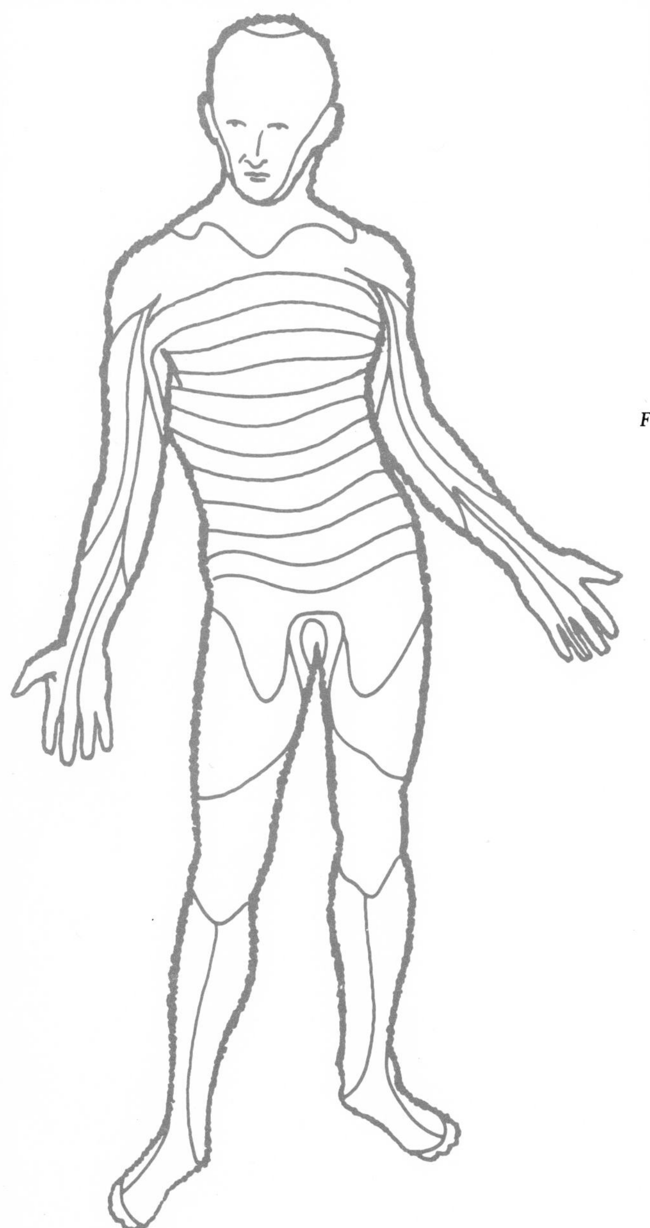

FOURTH EDITION

Structure and function of the body

Catherine Parker Anthony, R.N., B.A., M.S.

Formerly Assistant Professor of Nursing, Science Department, and Assistant Instructor of Anatomy and Physiology, Frances Payne Bolton School of Nursing, Case Western Reserve University, Cleveland, Ohio; formerly Instructor of Anatomy and Physiology, Lutheran Hospital and St. Luke's Hospital, Cleveland, Ohio

With 102 illustrations

The C. V. Mosby Company

Saint Louis 1972

Fourth edition

Second printing

Previous editions copyrighted 1960, 1964, 1968
Printed in the United States of America
International Standard Book Number 0-8016-0276-9
Library of Congress Catalog Card Number 77-187171
Distributed in Great Britain by Henry Kimpton, London

Preface

This fourth edition of *Structure and Function of the Body* aims at the same two goals as its predecessor. It seeks to help teachers teach, and practical nursing students and others learn basic information about the human body in an effective, efficient, and enjoyable manner. To try to achieve these goals, many features of preceding editions have been retained—features which so many of you have told the publisher and myself you found helpful. Introductory topical outlines, concluding summarizing outlines, and review questions for each chapter are among these. Illustrations, uncluttered by excessive details, and summarizing tables help make both students' and teachers' work easier. Again I have tried to breathe life into facts by citing many concrete examples and relating numerous data to medical and nursing procedures and laboratory tests.

Many changes appear in the book's content. Some, perhaps most, of these have stemmed from suggestions generously offered by instructors and students who used the third edition. Here are a few of the changes you can find: discussions of homeostasis, of the conduction system of the heart, and of the more common diseases related to body systems; additional information about skin functions, autonomic functions, hormones, liver functions, fetal circulation, control of the menstrual cycle, responses of the body to stress, and changes characteristic of aging. This edition also contains considerable new material. For instance, you can read about the increasingly popular procedure of vasectomy and about other information related to birth control. Another new feature is the listing of the terminology and abbreviations introduced in each chapter, which appears immediately before the outline summary of that chapter. The list of suggested supplementary readings (following the last chapter) has been made more helpful by organizing it according to chapters. Many timely, informative, and interesting articles have been added to this list.

Mr. Ernest W. Beck, an illustrious medical artist, has contributed excellent illustrations to this edition. More than twenty illustrations in all have been added.

Catherine P. Anthony

Contents

unit one

The body as a whole

unit two

Systems that form the framework of the body and move it

unit three

Systems that control body functions

unit four

Systems that process and distribute foods and eliminate wastes

unit five

Systems that reproduce the body

Anthony

Structure and function of the body

Fourth edition

unit one

The body as a whole

1

An introduction to the structure and function of the body

Have you ever asked yourself any questions like these—How do I stand erect and tall? How do I move? How do I do such simple, taken-for-granted things as washing my face, or brushing my teeth, or combing my hair? And how do diseases as different as poliomyelitis, strokes, and arthritis make me unable to move normally? If I touch something, how do I know that I have touched it and how do I detect whether it is hot or cold, smooth or rough, wet or dry? How do I see? Hear? Taste? Smell? Why must I eat and breathe to live? Why must my heart beat? Answers to these and many more questions can be found within the pages of this book. Chapter 1 begins with some general information about the structure of the body. It continues with a discussion of the body's main structural units and concludes with some general information about its functions.

General information about body structure

Cavities

One way that you can think of the body is as a building with only four rooms. The "rooms" are called cavities. To discover their names and locations, look at Figure 1-1. What anatomical name do you find for the space that you perhaps think of as your chest cavity? The figure identifies an abdominal cavity and a pelvic cavity. Actually they form only one cavity—the *abdominopelvic cavity*—since no partition of any kind separates them. Notice, however, that a partition does separate the thoracic cavity from the abdominal cavity. It is a dome-shaped muscle and is the most important muscle we have for breathing. What is its name? The space inside the skull is called the *cranial cavity*. It extends down inside the spinal column as the *spinal cavity*. The cranial and spinal cavities are classified as *dorsal cavities,* whereas the thoracic and abdominopelvic cavities are called *ventral cavities.* Dorsal, or posterior,

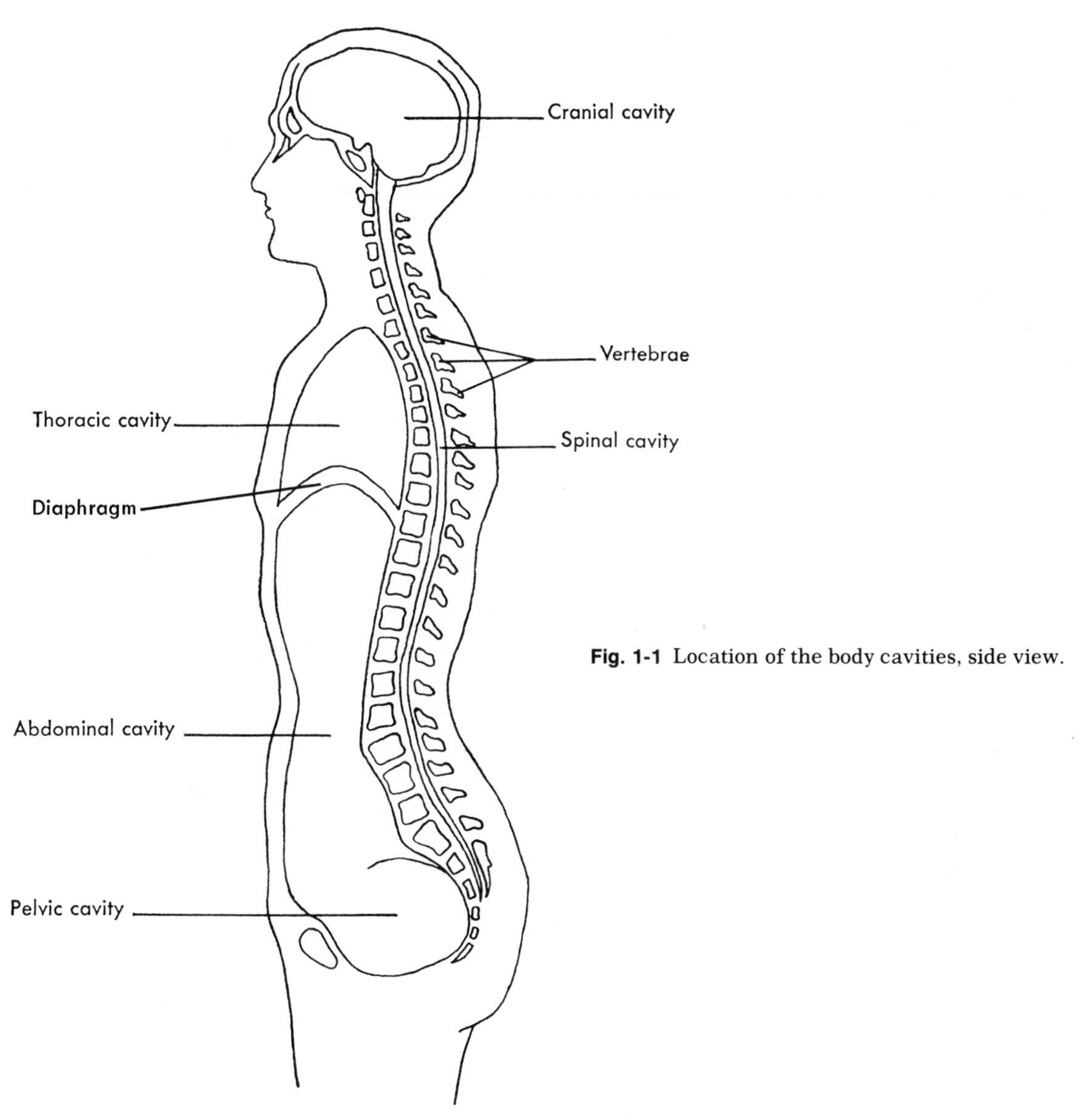

Fig. 1-1 Location of the body cavities, side view.

means back. Ventral, or anterior, means front.

Figure 1-2 shows some of the organs contained in the largest body cavities. In the thoracic cavity, for example, it shows the lungs, heart, and aorta. It does not show the various other vessels or nerves found in this cavity, or the thymus gland. In the abdominal cavity, it shows the small intestine and parts of the large intestine (the cecum and the ascending, transverse, and descending portions of the colon). Figure 1-2 does not show the gallbladder, the pancreas, the spleen, or the vessels and nerves that the abdominal cavity also contains.

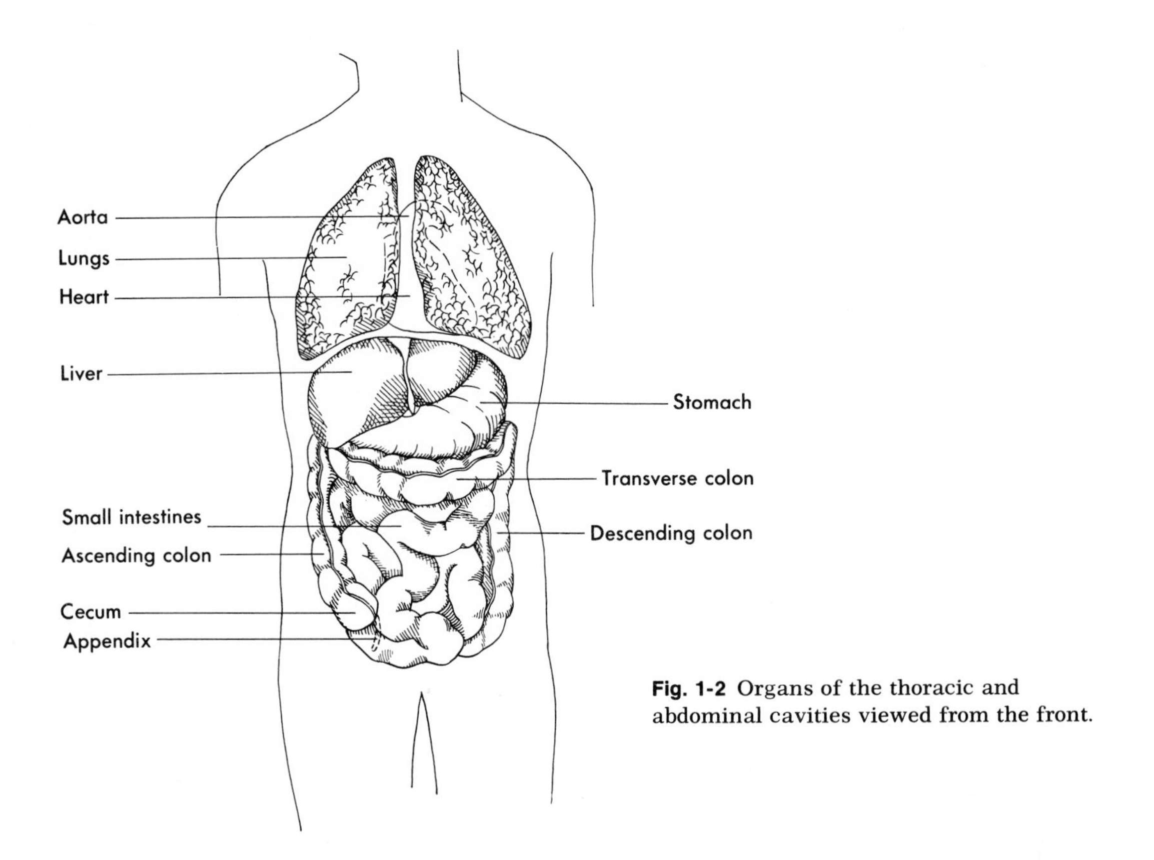

Fig. 1-2 Organs of the thoracic and abdominal cavities viewed from the front.

The two kidneys lie behind the abdominal cavity,* right underneath its lining. The reproductive organs, urinary bladder, and lowest part of the intestine fill the pelvic cavity. The brain occupies the cranial cavity. Find each body cavity in a model of the human body if you have access to one. Try to identify the organs in each cavity, and try to visualize their locations in your own body. Study Figures 1-1 and 1-2.

Structural units

Just as any building is made up of many kinds of structural units (walls, floors, steel, glass, bricks, nails, for instance), so too the body is made up of different kinds of structural units.

Here, listed in order from the smallest and simplest to the largest and most complex, are the names of the body's structural units: cells, tissues, organs, and systems.

About three hundred years ago, Robert Hooke looked through his microscope (one of the very early ones) at some plant material. What he saw must have surprised him. Instead of a single, much enlarged piece of plant material, he saw a group of many small pieces. They looked like miniature prison cells to him, so that is what he called them—cells. Hooke's discovery that plants were made up of cells proved to be a foundation stone for modern biol-

*A membrane called the *peritoneum* lines the abdominal cavity (see p. 90), and one called the *pleura* lines the thoracic cavity (discussed on p. 107).

ogy. Thousands of individuals have examined thousands of plant and animal specimens since Hooke's time and have found all of them made up of cells. Today, therefore, biologists think of a cell as the unit of structure of living things, just as you might think of a brick as the unit of structure of a brick wall or of a brick house. In other words, you might think of cells as the smallest "building blocks" of living things. Some living things are so simple that they are made up of just one cell; germs, for example (or to use the scientific term, microorganisms), consist of only one cell. Some living things are so complex that they consist of billions of cells – the human body, for example.

When a living thing is made up of a great many cells, its cells are not all alike. They differ somewhat in structure. Also, because structure determines function, they differ in certain functions. Cells with one type of structure, for example, conduct impulses; cells with a different type of structure contract; cells with still another type of structure secrete, and so on. Four main types of cells compose our bodies; epithelial cells, connective cells, muscle cells, and nerve cells.

A *tissue* is a group of similar cells with varying amounts and kinds of material filling in any spaces between the cells. Although there are only four main kinds of tissues – epithelial, connective, muscle, and nerve – there are many subtypes.

An *organ* is a structure composed of several kinds of tissues – often of all four main kinds. Each of the tissues in an organ in turn is composed of thousands of cells. Organs perform more complex functions for the body than any single cell or single tissue can perform alone. You probably already know the names of most of the organs – stomach, intestines, liver, gallbladder, heart, lungs, eyes, ears, and many others (Figure 1-3).

A *system* is the largest structural unit in the body. It consists of a group of organs that work together to perform a more complex function than any one organ can perform alone. Most anatomists group organs into the following nine systems: skeletal, muscular, circulatory, digestive, respiratory, urinary, reproductive, endocrine, and nervous (Figure 1-3).

What structures make up the body? Briefly, cells, tissues, organs, and systems. Millions of similar cells form each tissue, two or more kinds of tissues form each organ, several organs form each system, and nine systems form the body. Next we shall give a somewhat detailed description of cells and tissues. Organs and systems will be discussed in later chapters.

CELLS

Cell structure and cell function. Cells, the "building blocks" of the body, are extremely tiny. Most of them measure only 1/2,000 to 1/1,000 of an inch across. Think of that – living units so small that 2,000 of them lined up would form a row only an inch long! Obviously you cannot see anything that tiny with your naked eye. Cells are not macroscopic (visible to the naked eye), they are microscopic; only when they are magnified by a microscope can human cells be seen.

Cells differ in shape as well as in size. Some cells resemble tiny bricks; some are flat like the scales of a fish; some are long and slender like bits of thread; some are irregular in shape.

Cells have three main parts: cell membrane, cytoplasm, and nucleus. As you can see in Figure 1-4, the *cell membrane* forms the boundary of the cell. *Cytoplasm* makes up the bulk of its interior. The *nucleus* is a round or oval structure in the central part of the cell.

In addition to membrane, cytoplasm, and nucleus, cells also have almost countless smaller parts. Many of them are invisible even with the highest magnification a light microscope can provide (about a thousand times actual size). Light microscopes, therefore, did not reveal their existence, but electron microscopes did. With their magnifying power of many thousands of times, electron microscopes finally brought into man's view great numbers of ultra-small cell structures never seen before. And now, not many years after their discovery, some of their names—mitochondria, ribosomes, and lysosomes, for instance—have become familiar to almost everyone. Popular magazines and even newspapers frequently carry articles about them.

Everything known so far about these structures sounds fantastic—more like a science fiction writer's dream than a scientist's facts. For one thing, despite their incredibly small size, they each have definite structure. And a fact even harder to imagine—they carry on chemical functions at least as complicated as any performed by our most elaborate, modern chemical factories. These "little organs," or *organelles,* as they are called, serve the cell much as organs serve the body. Brief descriptions of the cell membrane, various organelles, and the nucleus follow.

Suppose you were to look at a cell magnified a thousand times by a light microscope. What could you observe about the structure of its *cell membrane*? Nothing more than its extreme thinness. Painstaking research has shown that the cell membrane measures only about 3/10,000,000 of an inch thick! But despite its fantastic delicacy, the cell membrane performs vital functions. It keeps the cell whole and intact, and it serves as a well-guarded gateway to the interior of the cell. It allows certain substances to move through it, but bars the passage of others. If it can no longer carry on these functions, the cell soon sickens and dies.

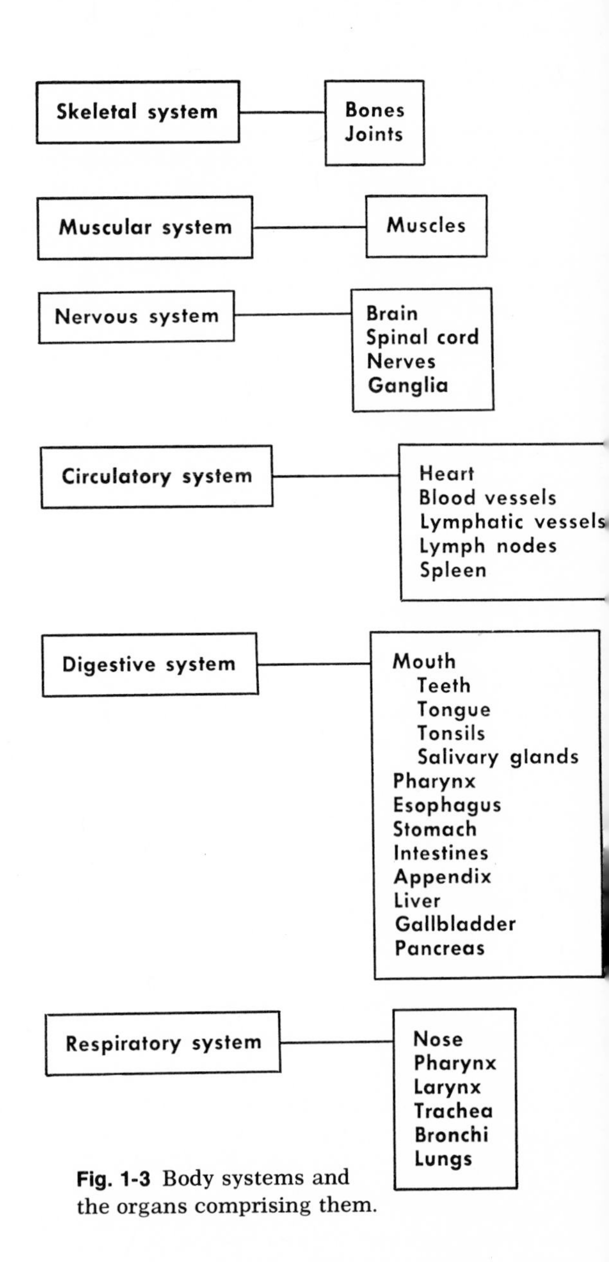

Fig. 1-3 Body systems and the organs comprising them.

Mitochondria are sausage-shaped structures so short that fifteen thousand or more of them could fit in a space just one inch long. Each mitochondrion, small as it is, consists of not one but two thin-membraned sacs, one inside the other. Observe their appearance in Figure 1-4. The life of every one of your body's cells depends upon the continuous functioning of these

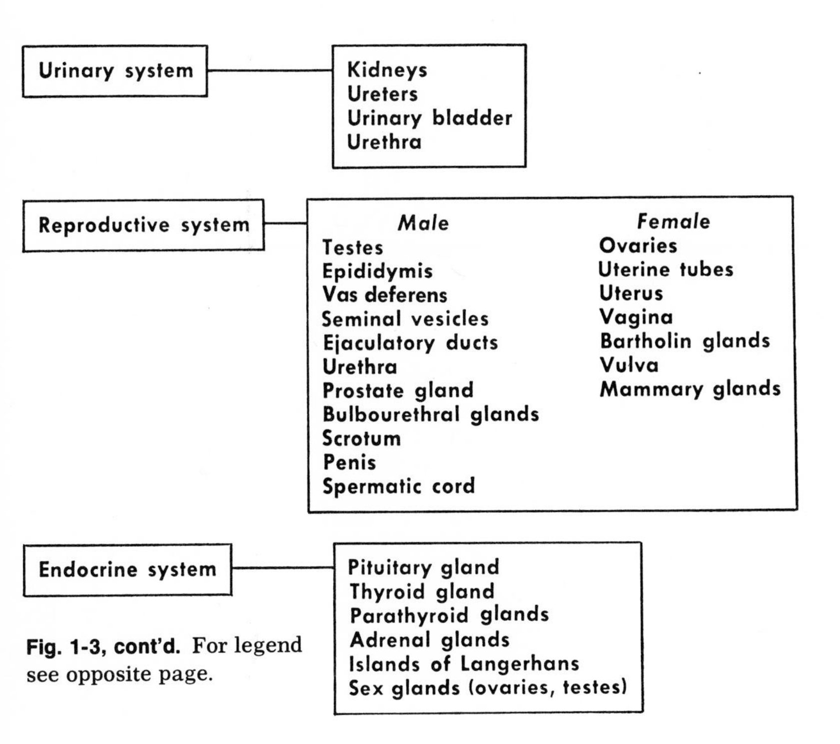

Fig. 1-3, cont'd. For legend see opposite page.

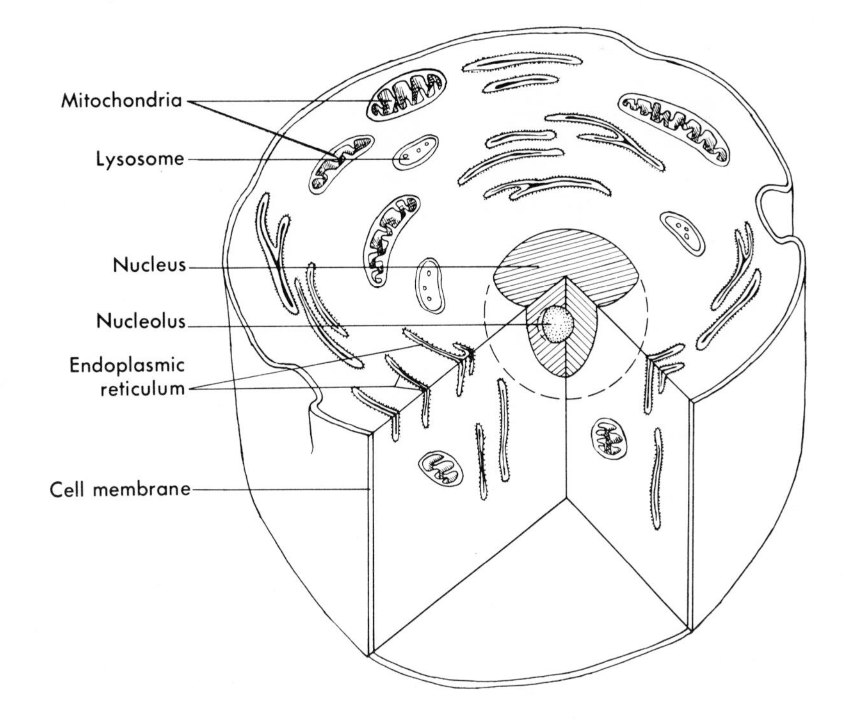

Fig. 1-4 Diagram of typical cell structure. Pie-shaped cut into the cell shows that cells are tridimensional, having thickness as well as length and width.

insignificant looking structures. Why? Because they serve as "power houses" for cells. By this we mean that they supply most of the energy cells use for doing the work that keeps them alive. Numerous complex chemical reactions go on within the mitochondria. These reactions make up the major part of a process called catabolism. Every cell uses food (metabolizes it) in two ways: it releases energy from some food molecules, and from others it makes a number of complex compounds—enzymes and hormones, for instance. *Catabolism* is the process by which cells release energy from foods. *Anabolism* is the process by which cells make more complex compounds from foods. *Metabolism* is the use cells make of foods; in other words, the process of metabolism consists of the two processes, catabolism and anabolism.

The main food catabolized by human cells is glucose, a sugar formed by the digestion of carbohydrate foods (starches and sugars). Most of the chemical reactions of catabolism take place in the mitochondria. Located in their membranous walls are the enzyme* molecules that catalyze these reactions. They change glucose into first one compound and then another, and finally, by adding oxygen they convert it to carbon dioxide and water. While these chemical changes are taking place the energy stored in the glucose molecule is being released. Almost instantaneously, however, more than half of it is put back into storage—but in the molecules of another compound, one popularly known as ATP (adenosine triphosphate). The rest of the energy originally stored in the glucose molecule is released as heat by the process of catabolism. ATP serves as the direct source of the energy for doing cellular work in all kinds of living organisms from one-celled plants to billion-celled animals, including man. Among biological compounds, therefore, ATP ranks as one of the most important. The energy in an ATP molecule differs from that in a food molecule in two ways: it can be released almost instantaneously, and as released, it can be used directly to do cellular work. Neither of these qualities characterizes the energy stored in glucose and other food molecules. Release of their energy occurs much more slowly, achieved only by a long series of chemical reactions, those that make up the process of catabolism. And for some reason such energy cannot be used directly for doing cellular work. It must first be transferred to ATP molecules and be released explosively from them. See Figure 1-5 for a summary of these

*An *enzyme* is a protein compound that has the special ability to speed up one particular chemical reaction. Every cell contains hundreds of different enzymes.

Fig. 1-5 A brief summary of glucose catabolism. (1) Series of chemical reactions, catalyzed by enzymes; occur in cytoplasm of cell; do not use oxygen. (2) Series of chemical reactions, catalyzed by enzymes; occur in mitochondria; use oxygen.

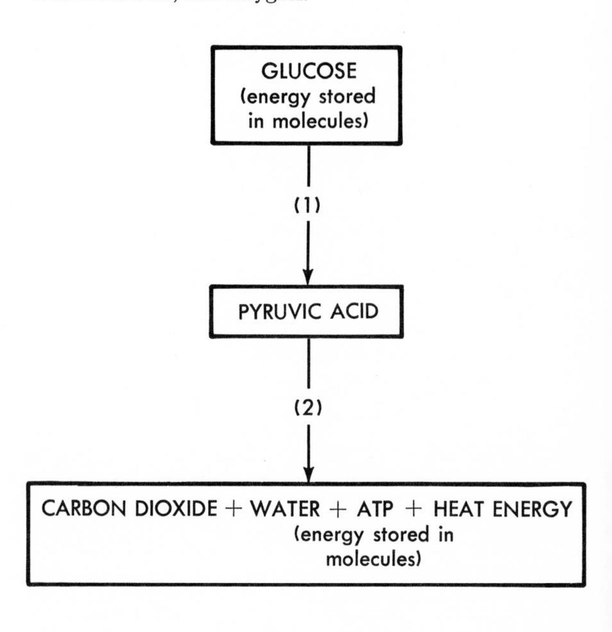

changes. Now before we leave the subject of mitochondria, we ought to answer a question asked at the beginning of this chapter, "Why do we have to breathe to stay alive?" The answer is that breathing is the only way we have for taking oxygen into our bodies. And only by using oxygen can our billions of mitochondria make available enough energy in a form that cells can use to do the work that keeps them, and us, alive.

Lysosomes are membranous-walled organelles, which in their active stage look like small sacs, often with tiny particles in them (Figure 1-4). Because lysosomes contain chemicals (enzymes) that can digest food compounds, one of their nicknames is "digestive bags." Lysosomal enzymes can also digest substances other than foods. For example, they can digest and thereby destroy microbes that manage to invade the cell. Thus, lysosomes can protect cells against destruction by microbes. But, paradoxically, they can also kill cells. If their powerful enzymes escape from the lysosome sacs into the cytoplasm, they destroy the entire cell by digesting it. This fact earned lysosomes their other nickname—"suicide bags."

Ribosomes, as seen under the powerful magnification of an electron microscope, look like small dots. Every cell contains thousands of them. You can see some of them in Figure 1-4 bordering the endoplasmic reticulum. Ribosomes carry on a most complex function, that of making enzymes and other protein compounds. This function earned ribosomes their nickname of "cellular protein factories."

The *endoplasmic reticulum* consists of a network of extremely small, extremely thin-walled canals. They follow a rather twisted path through the cell's cytoplasm and eventually open on its surface. Great numbers of ribosomes lie along the endoplasmic reticulum, giving it the dotted appearance shown in Figure 1-4. Some of the proteins made by ribosomes enter the canals of the endoplasmic reticulum and travel through them to other parts of the cell. In this respect, the endoplasmic reticulum functions for the cell as the circulatory system functions for the body.

The *Golgi apparatus,* long a mystery organelle, consists of tiny sacs stacked one upon the other near the nucleus. Only recently have scientists discovered what the Golgi apparatus does. It makes certain carbohydrate compounds, combines them with certain protein molecules, and packages the product in neat little globules. Then slowly these globules move outward to and through the cell membrane. Once outside the cell they break open, spilling their contents. An example of a Golgi apparatus product is the slippery substance known to all of us as mucus. If we wanted to nickname the Golgi apparatus, we might call it the cell's "carbohydrate producing and packaging" factory.

The *nucleus* of a cell, viewed under a light microscope, looks like a very simple structure indeed—just a small sphere in the central portion of the cell. In a suitably stained cell, the nucleus appears darker than the cytoplasm around it. In some cases, one can distinguish dark granules within the nucleus. These are the cell's *chromosomes,* whose small size greatly belies their importance. The main chemical composing them is beyond a doubt one of the most important of the thousands of compounds in the world. Surely, no other substance is more fascinating or more awesome. Even the name of this wondrous material is appropriately impressive—deoxyribonucleic acid, or DNA, for short. How can we justify making such extravagant claims for DNA? What is its function? In a single word, heredity. DNA passes on the characteristics of one generation of cells to the next. And, by so doing, it transmits from each generation of parents to their children all the traits, both physical and mental, that they inherit from their forebears. Stating this function is a simple and easy matter. But explaining it requires the telling of a long, complicated, and still unfinished story. We shall attempt only a brief synopsis.

The main theme of the DNA story revolves about the structure of the DNA molecule. The principle that structure determines function holds true for DNA as it does for all things, from molecules to machines, from a single cell to a human body. (The structure of a sewing machine, for instance, makes it possible for you to sew with it, but not to drive down a highway. To perform that function, quite obviously you must have a machine with an entirely different structure.)

DNA molecules have a strange and unique structure. To try to visualize it, picture first a long, long ladder. Imagine each of its rungs consisting of only two molecules. One may be a molecule of a compound named adenine. If it is, the other molecule joined to it is that of a compound called thymine. Only one other pair of molecules ever forms a rung of a DNA molecule. Their names are cytosine and guanine. Although DNA is an acid, all four of these substances—adenine, thymine, cytosine, and guanine—belong to the class of chemicals we call bases or alkalis. Adenine joined to thymine is called a *base-pair.* Cytosine joined to guanine is also called a base-pair. And although every rung of every DNA molecule consists of one or the other of these same two base-pairs, the order in which they follow one another differs in different DNA molecules. This fact about DNA structure is far from a meaningless detail. It has tremendous functional importance since the sequence of base-pairs is what determines heredity. In fact, that is what a *gene* is—a sequence of about one thousand base-pairs. And approximately 175,000 genes, cell specialists now tell us, compose the DNA molecule of a single human chromosome. By doing some simple multiplication, we arrive at some staggering figures:

Genes in 1 human chromosome = 175,000
Base-pairs in 1 gene = 1,000
175,000 × 1,000 = 175,000,000
Base-pairs in 1 human chromosome = 175,000,000

Genes in 1 human chromosome = 175,000
Chromosomes in one human cell = 46
175,000 × 46 = 8,050,000
Genes in 1 human cell = 8,050,000

It is little wonder, then, that no one of us inherits exactly the same traits as anyone else, with some 8,000,000 genes, or "heredity-bearers," in each of our cells.

Figure 1-6 illustrates the fact that base-pairs, either adenine-thymine, or cytosine-guanine compose the rungs of the DNA ladder. Examine this same figure to discover what kinds of compounds compose the sides of the ladder. The base-pairs attach to a sugar compound named what? What chemical lies between each two sugar molecules?

The DNA molecule has an intriguing shape. It is not a straight ladder, but a tortuously twisted one, actually more like a spiral staircase than a ladder. The title "world's longest spiral staircase" most surely belongs to the DNA molecule—some 175,000,000 steps and thousands upon thousands of turns long—and all of this fitted neatly into a microscopic size!

How do genes bring about heredity? There is, of course, no short or easy way to answer that question. In general, genes act in a highly complicated way to tell ribosomes what enzymes they are to make. According to the popular "one gene—one enzyme" theory, each gene—or in other words, each sequence in a DNA molecule of a thousand or so base pairs—is a code indicating the structure of one specific enzyme. This information is relayed to the

Table 1-1. Functions of some cell structures

Cell structures	*Main functions*
Cell membrane	Controls entrance and exit of substances into and out of cell; maintains cell's wholeness
Mitochondria	Catabolism
Lysosomes	Digestion
Ribosomes	Anabolism; make protein compounds
Endoplasmic reticulum	Transportation
Golgi apparatus	Makes carbohydrates, combines them with proteins, and packages product
Nucleus	Heredity

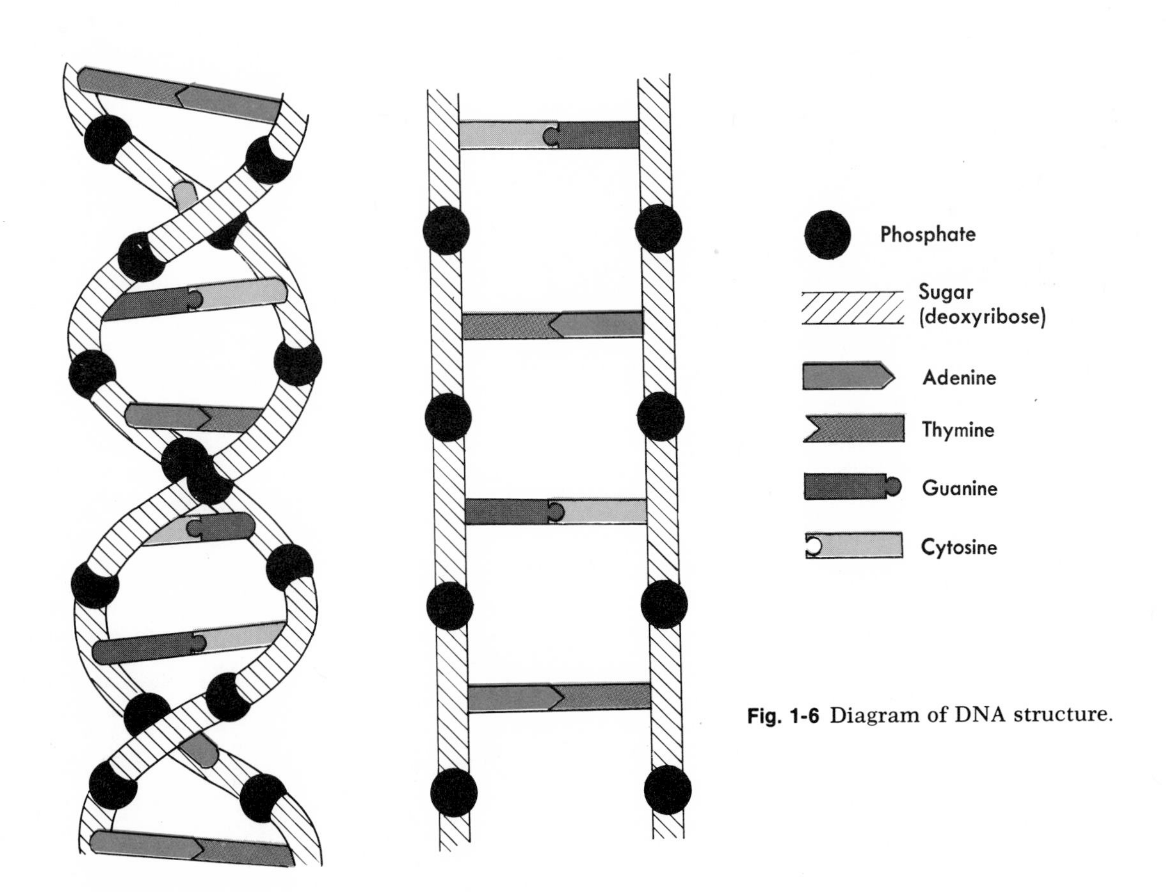

Fig. 1-6 Diagram of DNA structure.

ribosomes, and these cellular protein factories then make the enzymes. Enzymes are vital substances. Their job is to keep innumerable chemical reactions going on at the pace necessary for cells, and therefore the body, to stay alive. Summarizing, genes control enzyme production, enzymes control cellular chemical reactions, and cellular chemical reactions determine both cell structure and function and, therefore, heredity.

Tissues

Epithelial tissue. Epithelial tissue is especially good at performing three functions for the body—protection, absorption, and secretion. The reason that it is good at these functions is because of its structure. Epithelial tissue is a good example of the principle we stated earlier that stucture determines function.

There are several types of epithelial tissue. The structure of each one differs just enough from the structure of the others to make each type a specialist at a slightly different function. For example, in some parts of the body epithelial tissue consists of cells packed close together and arranged in several layers, one upon the other (Figure 1-7). This arrangement of the cells makes this type of epithelial tissue—called *stratified squamous epithelium*—a specialist at protection. For instance, stratified

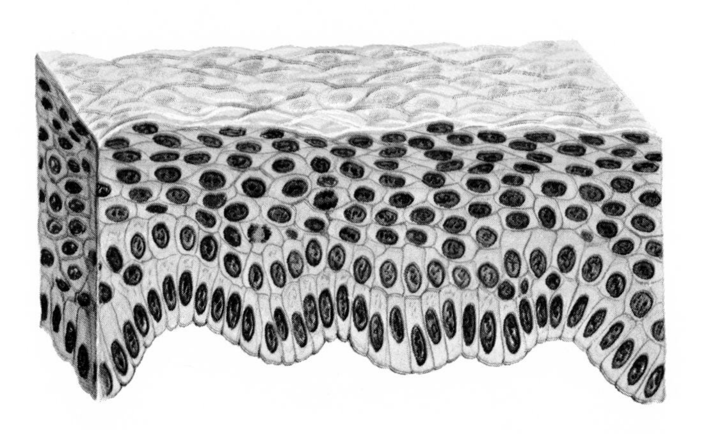

Fig. 1-7 Stratified squamous epithelium. Note the several layers of closely packed cells. Because of these structural features, stratified squamous epithelial tissue protects underlying structures against mechanical injury and microbe invasion.

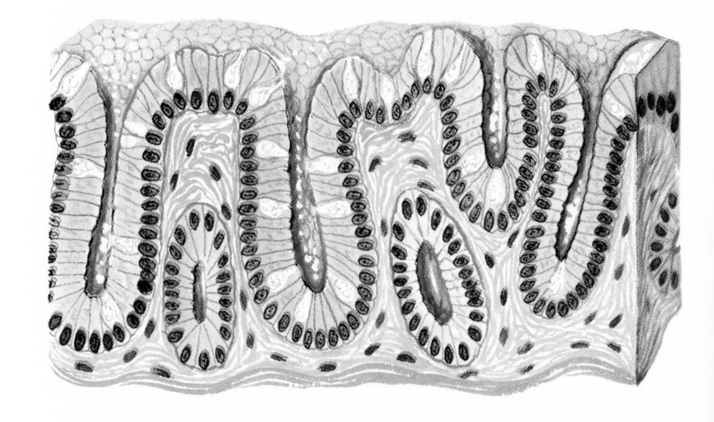

Fig. 1-8 Simple columnar epithelium with goblet cells, such as lines intestines. Secretion is the special function of this tissue.

squamous epithelial tissue protects the body against invasion by microorganisms. Most microorganisms cannot get through a barrier of stratified squamous tissue such as that which composes skin and mucous membrane surfaces. One way, therefore, that you can prevent infections is by taking good care of your skin. Don't let it become cracked from chapping. Guard against cuts and scratches—particularly good advice for nurses to remember.

In some parts of the body the main function of epithelial tissue is absorption. The cells of tissue in such areas are flat and arranged in a single layer. Substances therefore can pass through this tissue rapidly. This type of tissue is called *simple squamous epithelium.* (Squamous means shaped like fish scales.) It is the kind of tissue that forms the tiny air sacs in the lungs where oxygen is absorbed into the blood.

Secretion, the third function performed by epithelial tissue, is the chief function performed by a type of epithelial tissue called *simple columnar epithelium* (Figure 1-8).

Connective tissue. Connective tissue gets its name from its main function, that of connecting one kind of tissue to another. Most types of connective tissue also furnish support for some structure. There are many different types of connective tissue. Areolar, fibrous, adipose, cartilage, bone, and hemopoietic tissue are some of the main ones. Of these, areolar and fibrous are the most abundant. Areolar tissue serves as the "glue" of the body, so it is found all over the body wherever one kind of tissue attaches to another. Fibrous tissue also connects various structures. For instance, it holds bones together at the joints and anchors muscles firmly to bones. Hemopoietic tissue is a highly important type of connective tissue. It forms the various kinds of blood cells, and these in turn perform several vital functions.

The structure of connective tissue differs from that of other tissues primarily in the amount and kind of material found between its cells (in the amount of intercellular substance, that is) and in the many different varieties of its cells. Fibroblasts, macrophages, plasma cells, osteocytes, and fat cells are the names of a few of these different types of connective tissue cells. Figure 1-9 illustrates fat cells.

Muscle tissue. The special function of muscle tissue is contraction. There are three different types of muscle tissue: the tissue that makes up the muscles that are attached to bones, the tissue that helps form the walls of many tube-shaped organs (such as blood vessels and intestines), and the tissue that composes the heart. For a detailed discussion, see pp. 92 and 121.

Nervous tissue. Nervous tissue cells are discussed on p. 53.

ORGANS. The skin has the largest surface area of any organ in the body and is the heaviest. The functions it performs are vital for survival. We have chosen, therefore, to make it the subject of our discussion of organs.

The skin. Although the skin of an average-sized adult may weigh some twenty pounds or more, it is only paper-thin in some places and not much thicker in the thickest places. Architecturally the skin is a marvel. Consider the incredible number of structures fitted into an area no bigger than your little fingernail: several dozen sweat glands, hundreds of nerve endings, yards of tiny blood vessels, numerous oil glands and hairs, and literally thousands of cells.

Epidermis and dermis. Two kinds of tissue compose the skin (Figure 1-10). Stratified epithelial tissue makes up the outer or surface layer known as the *epidermis;* connective tissue makes up the thicker

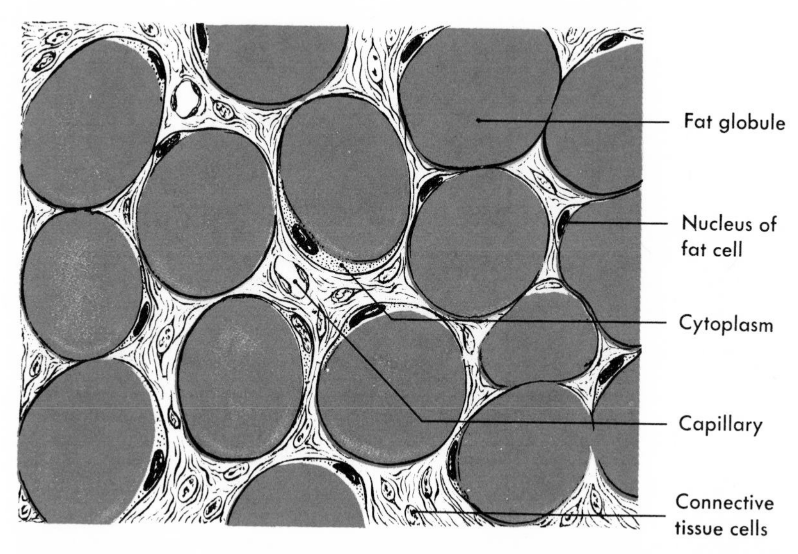

Fig. 1-9 Fat cells. In fat cells, which make up adipose tissue, a fat droplet occupies nearly the entire area of the cell, pushing the cytoplasm and nucleus out to the periphery of the cell.

underlying layer called the *dermis*. As you may recall, the cells of stratified epithelial tissue are closely packed together and are arranged in several layers. Only the cells of the innermost layer of the epidermis reproduce themselves. As they increase in number, they move up toward the surface and dislodge the dead horny cells of the outer layer. These flake off by the thousands onto our clothes, into our bath water, onto things we handle. Millions of epithelial cells reproduce daily to replace the millions shed—just one example of work our bodies do without our knowing it, even when they seem to be resting.

Have you ever wondered what gives color to your skin, or why its color differs from time to time? (At different seasons of the year, for example, or when you are sick, or angry, or frightened, or too warm, or too cold?) Skin color depends upon two factors: (1) the amount of a pigment, called melanin, in the epidermis; and (2) the amount of blood in the dermis and also the kind of blood (that is, whether the red blood cells are carrying much or little oxygen). Light-colored skin tans after long exposure to sunlight because ultraviolet rays increase the formation of melanin. Skin reddens when heat or emotions cause its capillaries to fill with more blood. It pales when sickness or cold decreases the amount of blood. It becomes bluish when the blood contains too little oxygen. Doctors and nurses use this knowledge to judge quickly whether a normal amount of blood is circulating through the skin and whether blood is carrying enough oxygen to body cells.

The deep layer of the skin, the dermis, has several structural points of interest. Look at the palms of your hands and you will see one of these—the ridges and grooves that make possible fingerprinting as a means of identification. Note in Figure 1-10 how the epidermis follows the contours of ridges present in the dermis.

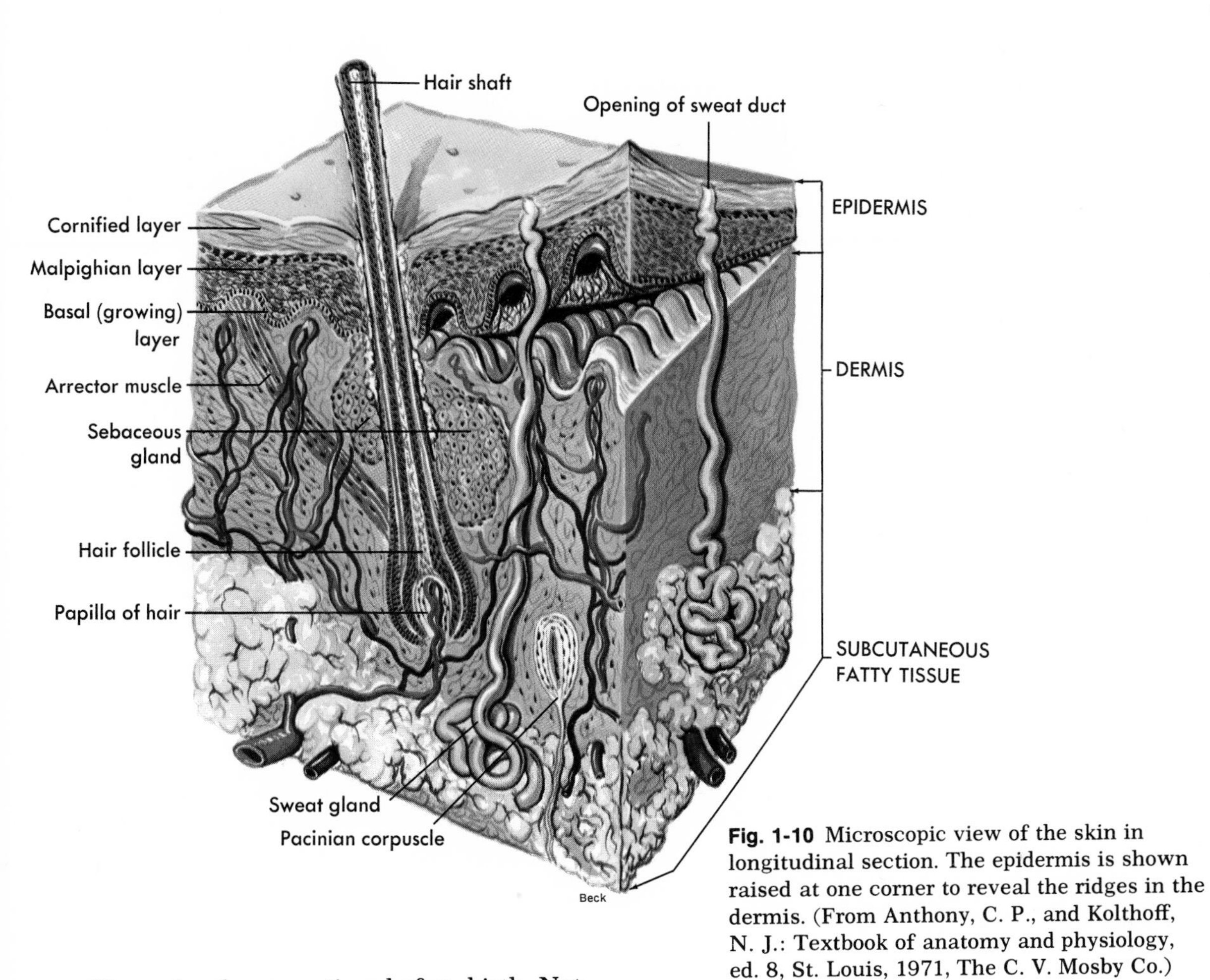

Fig. 1-10 Microscopic view of the skin in longitudinal section. The epidermis is shown raised at one corner to reveal the ridges in the dermis. (From Anthony, C. P., and Kolthoff, N. J.: Textbook of anatomy and physiology, ed. 8, St. Louis, 1971, The C. V. Mosby Co.)

These develop sometime before birth. Not only is their pattern unique for each individual, but also it never changes except to grow larger—two facts that explain why our fingerprints or footprints positively identify us. Many hospitals today identify each newborn baby by footprinting it almost as soon as it is born.

Connective tissue composes the dermis. Instead of cells being crowded close together like the epithelial cells of the epidermis, they are scattered far apart, with many fibers in between. Some of the fibers are tough and strong (collagenous or white fibers), and others are stretchable and elastic (elastic fibers). Some of the cells of the dermis store fat. The wrinkles that come with age are thought to be caused partly by a decrease in skin fat and partly by a decrease in the number of elastic fibers. A rich network of capillaries lies in the dermis and also immediately beneath it. The epidermis contains no blood vessels. Surgeons sometimes use the knowledge of this anatomical fact to treat certain skin conditions by cutting such thin shavings from the surface of the skin that they remove only the epidermis. As you would expect, this is a bloodless operation. Examine Figure 1-10 to find the answers to the following questions. Which layer of the skin—epidermis or dermis—contains hair follicles? Oil (sebaceous) glands? Sweat glands? Both the blood vessels and the nerves of the skin lie in its dermal layer.

Hair. If you have ever looked closely at a newborn baby, you probably noticed its soft downy hair. If the baby was born prematurely, you may have found the same type of hair (called "lanugo," from the Latin word that means down or wool) all over its tiny body. Coarser hair soon replaces the lanugo of the scalp and eyebrows, but new hair on the rest of the body generally stays delicate and downlike. Hair grows from a cluster of epithelial cells at the bottom of the hair follicle. As long as these cells remain alive, new hair will replace any that is cut or plucked. Contrary to popular belief, frequent cutting or shaving does not make hair grow faster or become coarser, because neither process affects the epithelial cells that form the hairs, since they are embedded in the scalp.

Oil glands. Wherever hairs grow, oil or sebaceous glands also grow. Their tiny ducts open into hair follicles so that their secretion (sebum) lubricates the hair as well as the skin. Someone aptly described sebum as "nature's cold cream." Besides keeping the hair and skin soft and pliant, this natural oil also helps preserve the normal water content of the body by hindering water loss from the skin by evaporation. This in turn helps preserve normal body temperature. Because evaporation has a cooling effect and because the presence of oil on the skin slows down evaporation, it also slows down cooling. Long-distance swimmers apply this knowledge when they coat their bodies with grease before plunging into the water.

A small involuntary muscle attaches to the side of each hair follicle just below its oil gland and extends upward on a slant to attach to the skin. When it contracts, as it often does when we are cold or emotionally upset, it produces several effects. It squeezes on the oil gland with the same result that you would produce if you squeezed on an oil can—a tiny bit of oil squirts out of each follicle onto the skin. Also, as the muscle contracts it simultaneously pulls on its two points of attachment—that is, up on a hair follicle but down on a point of skin. This produces little raised places (goose pimples) between the depressed points of the skin and, at the same time, pulls the hairs up more or less straight. The latter fact accounts for the name of these muscles, arrectores pilorum (Latin for "erectors of the hair"). We unconsciously recognize these facts in pictorial expressions such as "I was so frightened my hair stood on end" and "She was so angry she was fairly bristling."

Sweat glands. The pinpoint-sized openings on the skin that you probably think of as pores are really outlets of small ducts from the sweat glands. The glands themselves lie in the subcutaneous tissue just under the dermis, and their ducts spiral up through the epidermis to the surface. There are great numbers of them. For instance, the palms of your hands are estimated to have about 3,000 sweat glands per square inch! The soles of the feet, the forehead, and the axillae (armpits) also contain a great many of them, more than most parts.

The skin is a vital organ. So essential are its functions that if, for instance, a burn destroys about one-third of a healthy young adult's skin, he probably will die. The skin contributes the following major functions. It serves as our first line of defense against a multitude of hazards. It protects us against the daily invasion of millions of deadly microorganisms and chemicals and prevents the sun's ultraviolet rays from penetrating into the interior of our bodies. The skin's nerve endings inform us of our continually changing external environment. They tell us when something touches us or presses against us. They make us aware of heat and cold and pain. Changing amounts of sweat secreted by the skin's sweat glands help regulate both the body's temperature and its fluid content. Changes in the volume of blood flowing through the skin play a part in steadying both body temperature and blood pressure. Chapters 4, 5, and 6 give more information about how the skin performs some of these lifesaving functions.

SYSTEMS. The nine systems are discussed in the remaining chapters of this book.

General information about body function

The function of any body part is determined by its structure. Therefore, if the structure of a part changes markedly, its functioning also changes. Disease and injury frequently produce rapid changes in structure and function, whereas the passage of time brings about more gradual changes. Even casual observation reveals many differences between, for example, the bodies of a baby, an 8-year-old, a 20-year-old, and an 80-year-old.

All body functions are, in the last analysis, cell functions. Each cell performs not only self-serving functions to maintain its own life but also specializes in some body-serving function to help maintain the whole body's life. Muscle cells to mention one example, specialize in contraction, a function that serves the body by producing its movements including those that bring about breathing.

Cells must have a relatively steady environment in order to function normally and survive. But the cells live in an environment entirely different from the environment of the body. The body is a land dweller, living in the external environment of air. Body cells are water dwellers, living in the internal environment of fluid. The internal environment, the fluid around cells, has a name descriptive of its location —*extracellular fluid.* When the chemical composition, temperature, and volume of the extracellular fluid remain almost constant, fluctuating only slightly above or below a midpoint, the body is said to be maintaining a condition of *homeostasis.* Homeostasis is a requirement of number one importance for normal functioning of cells and healthy survival of the body. Many devices, called *homeostatic mechanisms,* function continually to maintain or quickly restore homeostasis. Most of the chapters in this book describe at least one important homeostatic mechanism.

Disease

According to one definition, disease is any abnormal condition of the body. It may involve a single part or the entire body. It may be an abnormality of either structure or function. More often, however, disease includes abnormalities of both structure and function. Although the main subject of this book is the normal or healthy body, we shall conclude each chapter with some brief remarks about diseases related particularly to the parts of the body discussed in that chapter. For this chapter, therefore, the diseases we shall mention represent abnormalities of cells and tissues.

Hereditary diseases are abnormalities transmitted by genes. Or, to use another expression, hereditary diseases are inborn abnormalities. Mongolism (or Down's syndrome) was the first such disease known. In 1959 investigators reported that the cells of mongoloid individuals contained an abnormal number of chromosomes—47 instead of 46. Since that time, scientists have discovered that abnormal genes or chromosomes cause a great many other diseases. PKU is one you may have heard of. (The initials stand for phenylketonuria.) A single "bad gene" produces PKU. The blood of a baby born with this disease lacks only one of the enzymes necessary for metabolizing a certain protein substance, but this one enzyme is extremely important. If the baby's food contains the substance normally acted on by this enzyme, in time the absence of this enzyme acting on the substance will result in brain damage and mental retardation. Today, however, physicians have available an easy test by which they can detect PKU early enough to prescribe a diet for the baby that will prevent any brain damage. Other well-known hereditary disorders are hemophilia and color blindness.

Diseases of the skin are many and varied. So much so, in fact, that some physicians (dermatologists) specialize in diagnosing and treating skin disorders.

outline summary

GENERAL INFORMATION ABOUT BODY STRUCTURE

1 Ventral cavities
  a Thoracic cavity
    Contains lungs, heart, trachea, esophagus, and thymus gland
  b Abdominopelvic cavity
    Abdominal cavity contains stomach, intestines, liver, gallbladder, pancreas, and spleen
    Pelvic cavity contains reproductive organs, urinary bladder, lowest part of intestine
2 Dorsal cavities
  a Cranial cavity
    Contains brain
  b Spinal cavity
    Contains spinal cord

STRUCTURAL UNITS

1 Cells—structural units of all living things
2 Tissues—groups of like cells
3 Organs—composed of more than one kind of tissues; perform more complex functions than a single tissue
4 Systems—groups of organs; perform more complex functions than a single organ

Cells

1 Structure
  a Size—microscopic but varies with type of cell
  b Shape—varies with type of cell
  c Parts—cell membrane, cytoplasm, organelles (mitochondria, lysosomes, ribosomes, endoplasmic reticulum, Golgi apparatus), nucleus (contains chromosomes, composed of DNA)
2 Functions
  a Each cell performs functions necessary for its own life, such as intake of oxygen and food and metabolism (catabolism plus anabolism)
  b Different cell structures perform different functions—see Table 1-1
  c Each cell performs some special function for body as a whole, such as contraction by muscle cells, conduction of impulses by nerve cells

Tissues

1 Epithelial tissue
  a Functions—protection, absorption, and secretion
  b Types—several; for example, stratified and simple squamous epithelial tissue
2 Connective tissue
  a Functions—connection, support, and blood cell formation
  b Types—numerous; for example, areolar, fibrous, adipose, cartilage, bone, blood, and hemopoietic tissues
3 Muscle tissue
  a Function—contraction
  b Types—discussed in Chapter 3
4 Nervous tissue—discussed in Chapter 4

Organs

Skin

1 Structure
  a Epidermis—outer layer of stratified squamous epithelial tissue; dead surface cells continually flake off; contains pigment (melanin)
  b Dermis—underlying layer of connective tissues; ridges and grooves in dermis form pattern unique to each individual—basis of fingerprinting; blood vessels, nerves, hair roots, oil glands, sweat glands; fat stored
  c Hair—grows from cluster of epithelial cells at bottom of hair follicle deep in dermis
  d Oil (sebaceous) glands—secrete oil into hair follicles; oil keeps hair and skin soft; also helps prevent loss of water and heat from skin
  e Sweat glands—openings from their ducts are "pores" of skin; amount of sweat secreted helps control amount of water and heat lost from body

2 Functions
   a Protection
   b Sensations
   c Helps regulate body temperature and blood pressure

GENERAL INFORMATION ABOUT BODY FUNCTION

1 Function is determined by structure; changes in structure bring about changes in function
2 Body functions are cell functions
3 Homeostasis, that is, a relatively steady internal environment, is essential for healthy survival

DISEASE

1 Hereditary diseases—abnormalities transmitted by genes
   a Mongolism (Down's syndrome)—47 chromosomes per cell instead of 46, the normal number
   b PKU—caused by one abnormal gene resulting in lack of one enzyme necessary for normal metabolism of one protein substance; leads to brain damage and mental retardation
   c Hemophilia, color blindness, and many other diseases
2 Skin diseases—great number of them; dermatology is branch of medicine specializing in diseases of skin

new words and abbreviations

anabolism	gene
ATP	Golgi apparatus
catabolism	lysosome
cell	metabolism
chromosome	mitochondria
cytoplasm	nucleus
dermis	organ
DNA	organelle
endoplasmic reticulum	ribosome
enzyme	system
epidermis	tissue

review questions

1 In what cavity could you find each of the following?

appendix	liver
brain	lungs
esophagus	pancreas
gallbladder	spinal cord
heart	spleen
intestines	urinary bladder

2 Explain what the following terms mean: cells, organs, systems, tissues.
3 What is the primary function of the cell membrane?
4 What and where are the following? What functions do they perform?

endoplasmic reticulum	lysosomes
	mitochondria
Golgi apparatus	ribosomes

5 What is the full name of the now famous acid found in the nuclei of cells?
6 This acid makes up the major part of what microscopic structures?
7 Very briefly, what function does DNA perform?
8 Give a modern definition of the word gene.
9 What functions does epithelial tissue specialize in? Muscle tissue? Connective tissue?
10 Describe briefly the epidermis and dermis.
11 Why is oil and sweat secretion by the skin glands important to the body?
12 Give the scientific name for the cellular structure indicated by each of the following nicknames: "carbohydrate producing and packaging factory," "digestive bags," "power houses," "protein factories," and "suicide bags."
13 In one word, what is the function of DNA?
14 About how many genes is one human cell estimated to contain?

unit two

Systems that form the framework of the body and move it

2

The skeletal system

Unit two consists of two chapters. Chapter 2 contains a discussion of the skeletal system, the system that provides the body with a rigid framework. In this respect the skeletal system functions as steel girders do in a building. But it also functions quite differently. Unlike steel girders, bones can move toward or away from each other and even, in some cases, can move around in circles. Chapter 3 deals with the system that moves bones.

Before studying this chapter, let your imagination go for a few minutes. Think of your bones suddenly turning soft, into a material, say, of the consistency of a piece of liver. Suppose you were standing when this change took place. What do you see happening? Suppose you fell and struck your head. Would you bruise your scalp only or your brain as well? Now change your mental picture. This time think of a single piece of wood cut in the shape of the human body. Try to visualize this structure sitting down, bending over, throwing a ball, or walking across a room. Can you see it moving any of its parts or making any one of the hundreds of movements that you make every day without even giving them a thought?

Functions

The images you have just called up point out three important functions of the skeletal system: it supports and gives shape to the body, it protects internal organs, and it makes movements possible.

How the skeletal system makes movements possible will be considered partly in this chapter in the discussion of joints and partly in the next chapter in the discussion of muscle actions. We shall consider first the general plan of the skeleton and individual bones.

General plan

The human skeleton has two divisions: the *axial skeleton* and the *appendicular skeleton*. Read Table 2-1 to find out what parts make up each of these two divisions.

Skull

Twenty-eight bones compose the skull. You will probably want to learn their names and find out what part of the skull each one forms. Their names are given in Table 2-2. Find as many of them as you can on Figure 2-1, and feel their outlines on your own body where possible. Examine them on a skeleton if you have access to one.

"My sinuses give me so much trouble." Have you ever heard this complaint or perhaps uttered it yourself? *Sinuses* are spaces, or cavities, inside some of the cranial bones. Four pairs of them (those in the frontal, maxillary, sphenoid, and ethmoid

Text continued on p. 28.

Table 2-1. Main parts of the skeleton

Axial skeleton	*Appendicular skeleton*
Skull	Upper extremities
Cranium	Shoulder girdle
Face	Arms
Spine	Hands
Vertebrae	Lower extremities
Thorax	Hip girdle
Ribs	Legs
Sternum	Feet
Ear bones	
Hyoid bone	

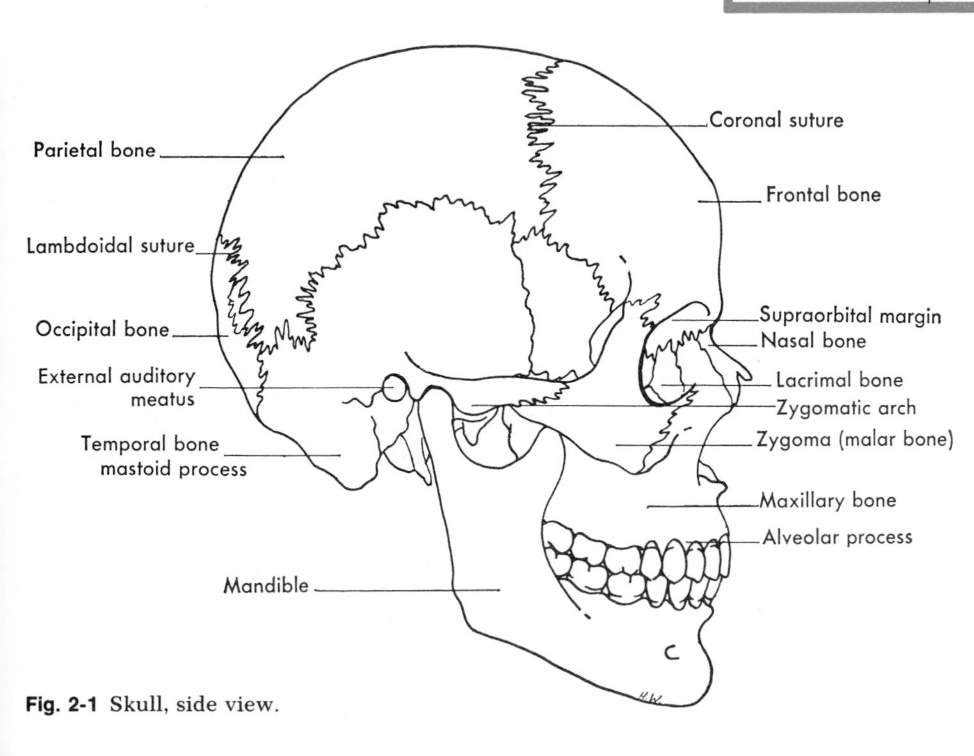

Fig. 2-1 Skull, side view.

Table 2-2. Bones of the skeleton

Name	*Number*	*Description*
Cranial bones		
Frontal	1	Forehead bone; also forms front part of floor of cranium and most of upper part of eye sockets; cavity inside bone above upper margins of eye sockets (orbits) called *frontal sinus;* lined with mucous membrane
Parietal	2	Form bulging topsides of cranium
Temporal	2	Form lower sides of cranium; contain *middle and inner ear structures; mastoid sinuses* are mucous-lined spaces in *mastoid process,* the protuberance behind ear; *external auditory canal* is tube leading into temporal bone
Occipital	1	Forms back of skull; spinal cord enters cranium through large hole *(foramen magnum)* in occipital bone
Sphenoid	1	Forms central part of floor of cranium; pituitary gland located in small depression in sphenoid called *Turk's saddle* or sella turcica
Ethmoid	1	Complicated bone that helps form floor of cranium, side walls and roof of nose and part of its middle partition (nasal septum), and part of orbit; contains honeycomblike spaces, the *ethmoid sinuses; superior and middle turbinate bones* (conchae) are projections of ethmoid bone; form "ledges" along side wall of each nasal cavity
Face bones		
Nasal	2	Small bones that form upper part of bridge of nose
Maxillary	2	Upper jawbones; also help form roof of mouth, floor, and side walls of nose and floor of orbit; large cavity in maxillary bone is *maxillary sinus*
Zygoma (malar)	2	Cheek bones; also help form orbit
Mandible	1	Lower jawbone
Lacrimal	2	Small bone; helps form medial wall of eye socket and side wall of nasal cavity
Palatine	2	Form back part of roof of mouth and floor and side walls of nose and part of floor of orbit
Inferior turbinate	2	Form curved "ledge" along inside of side wall of nose, below middle turbinate
Vomer	1	Forms lower, back part of nasal septum
Ear bones		
Malleus	2	Malleus, incus, and stapes are tiny bones in middle ear cavity in temporal bone; name, malleus, means hammer – shape of bone
Incus	2	Incus means anvil – shape of bone
Stapes	2	Stapes means stirrup – shape of bone
Hyoid bone	1	U-shaped bone in neck at base of tongue
Vertebral column		
Cervical vertebrae	7	Upper seven vertebrae, in neck region; first cervical vertebra called *atlas;* second called *axis*
Thoracic vertebrae	12	Next twelve vertebrae; ribs attach to these
Lumbar vertebrae	5	Next five vertebrae; those in small of back
Sacrum	1	In child, five separate vertebrae; in adult, fused into one
Coccyx	1	In child, three to five separate vertebrae; in adult, fused into one
Thorax		
True ribs	14	Upper seven pairs; attach to sternum by way of *costal cartilages*
False ribs	10	Lower five pairs; lowest two pairs do not attach to sternum, therefore, called *floating ribs;* next three pairs attach to sternum by way of costal cartilage of seventh ribs
Sternum	1	Breast bone; shaped like a dagger; piece of cartilage at lower end of bone called *xiphoid process*

Table 2-2. Bones of the skeleton—cont'd

Name	*Number*	*Description*
Upper extremities		
Clavicle	2	Collarbones; only joints between shoulder girdle and axial skeleton are those between each clavicle and sternum
Scapula	2	Shoulder bones; scapula plus clavicle forms *shoulder girdle; acromion process*—tip of shoulder that forms joint with clavicle; *glenoid cavity*—arm socket
Humerus	2	Upper arm bone
Radius	2	Bone on thumb side of lower arm
Ulna	2	Bone on little finger side of lower arm; *olecranon process*—projection of ulna known as elbow or "funny bone"
Carpal bones	16	Irregular bones at upper end of hand; anatomical wrist
Metacarpals	10	Form framework of palm of hand
Phalanges	28	Finger bones; three in each finger, two in each thumb
Lower extremities		
Pelvic bones	2	Hip bones; *ilium*—upper flaring part of pelvic bone; *ischium*—lower back part; *pubic bone*—lower front part; acetabulum—hip socket; *symphysis pubis*—joint in midline between two pubic bones; pelvic inlet—opening into *true pelvis,* or pelvic cavity; if pelvic inlet is misshapen or too small, infant skull cannot enter true pelvis for natural birth
Femur	2	Thigh or upper leg bones; *head of femur*—ball-shaped upper end of bone; fits into acetabulum
Patella	2	Kneecap
Tibia	2	Shin bone; *medial malleolus*—rounded projection at lower end of tibia commonly called inner ankle bone
Fibula	2	Long slender bone of lateral side of lower leg; *lateral malleolus*—rounded projection at lower end of fibula commonly called outer ankle bone
Tarsal bones	14	Form heel and back part of foot; anatomical ankle
Metatarsals	10	Form part of foot to which toes attach; tarsal and metatarsal bones so arranged that they form three arches in foot: *inner longitudinal arch* and *outer longitudinal arch,* both of which extend from front to back of foot, and transverse or *metatarsal arch* that extends across foot
Phalanges	28	Toe bones; three in each toe, two in each great toe
Total	206	

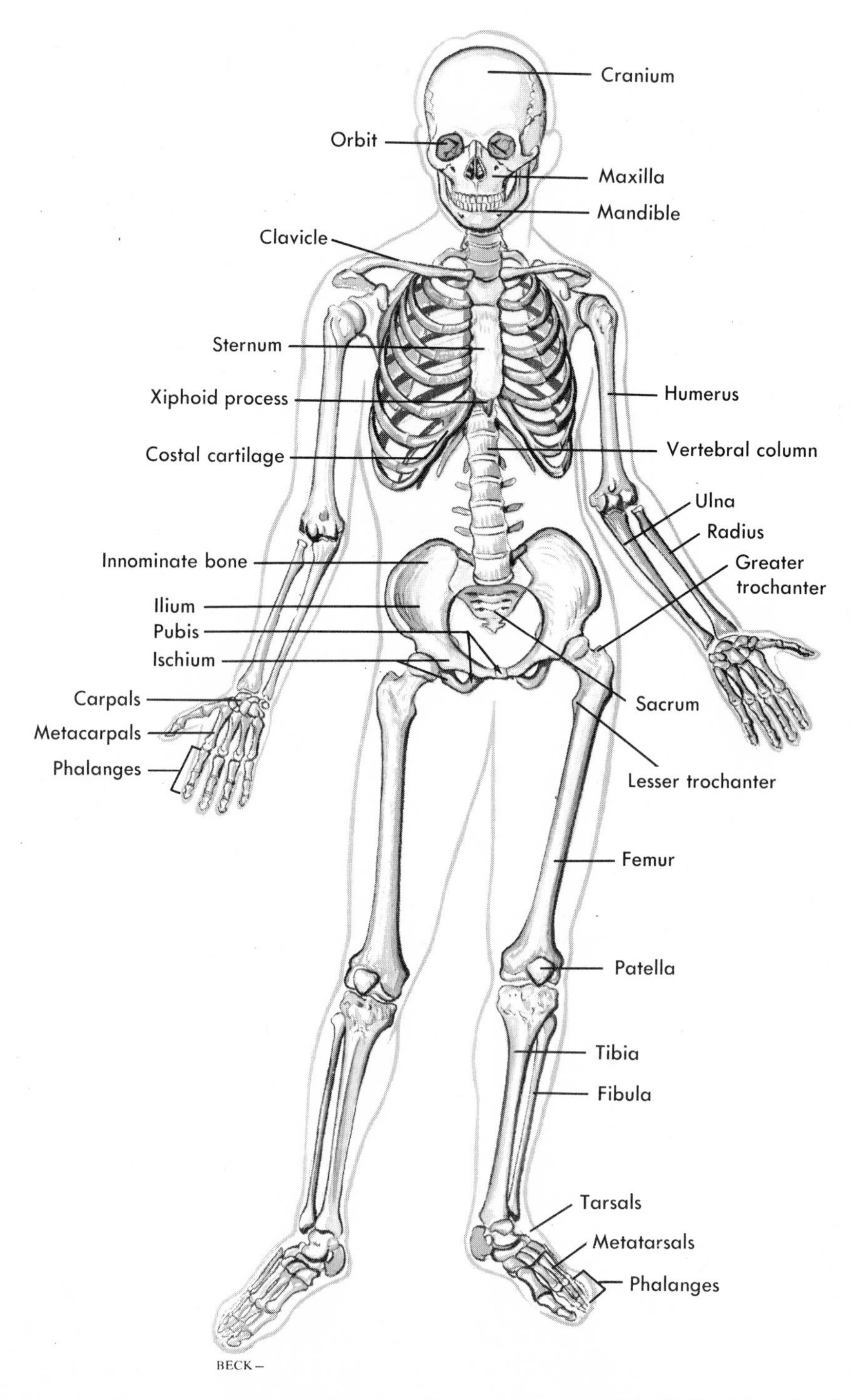

Fig. 2-2 Skeleton, anterior view. (From Anthony, C. P., and Kolthoff, N. J.: Textbook of anatomy and physiology, ed. 8, St. Louis, 1971, The C. V. Mosby Co.)

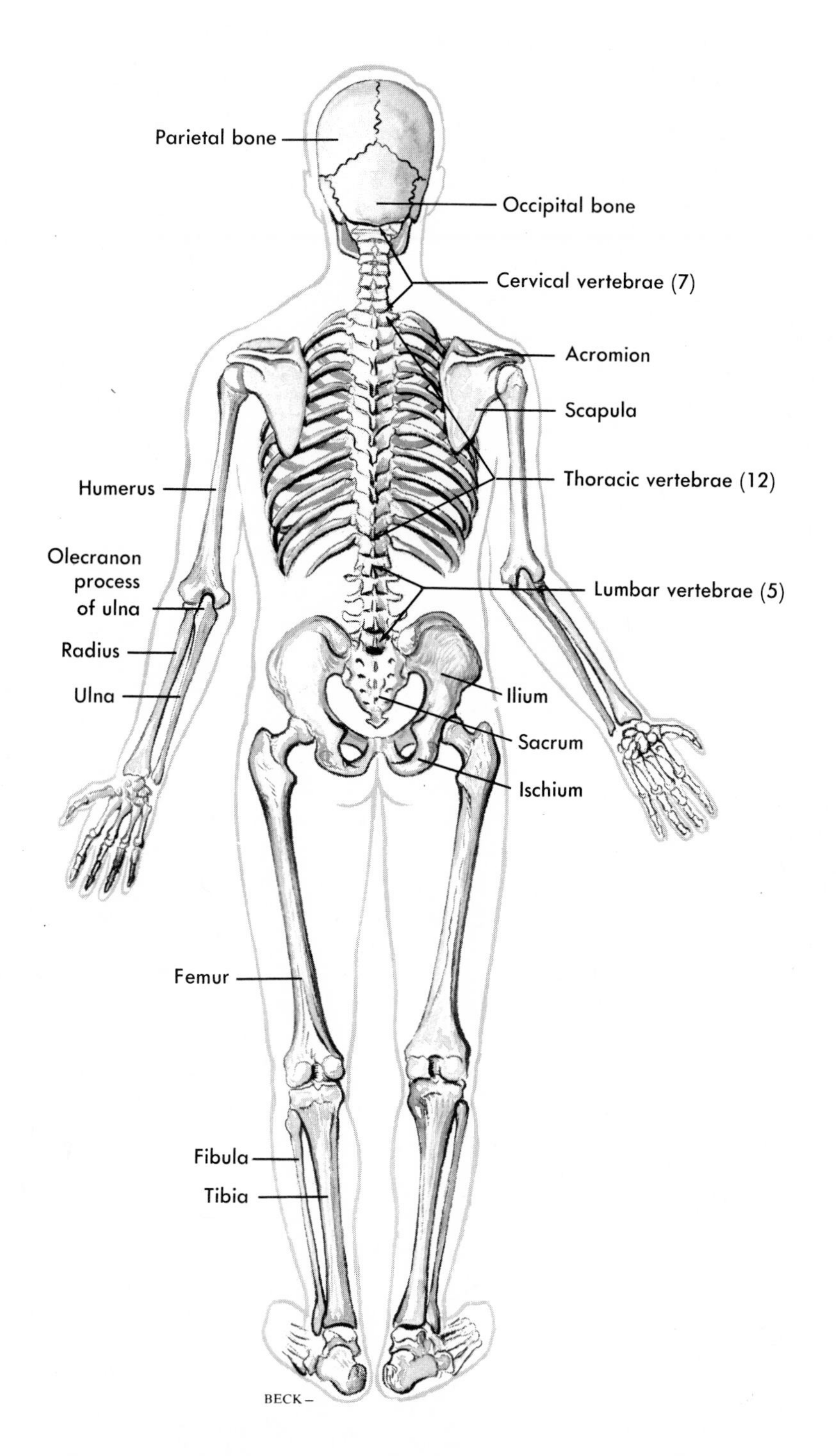

Fig. 2-3 Skeleton, posterior view. (From Anthony, C. P., and Kolthoff, N. J.: Textbook of anatomy and physiology, ed. 8, St. Louis, 1971, The C. V. Mosby Co.)

bones) have openings into the nose and so are referred to as *paranasal sinuses.* Sinuses give trouble when the mucous membrane that lines them becomes inflamed, swollen, and painful. For example, inflammation in the frontal sinus *(frontal sinusitis)* often starts from a common cold. (The letters "itis" added to a word mean inflammation of.)

Spine (vertebral column)

The term "vertebral column" may suggest a mental picture of the spine as a single long bone shaped like a column in a building, but this is far from true. The vertebral column consists of a series of separate bones (vertebrae) connected in such a way that they form a flexible curved rod. Different sections of the spine have different names: cervical region, thoracic region, lumbar region, sacrum, and coccyx. Read about them in Table 2-2.

Have you ever noticed the four curves in your spine? Your neck and the small of your back curve slightly inward or forward, whereas the chest region of the spine and the lowermost portion curve in the opposite direction. The cervical and lumbar curves of the spine are called concave curves, and the thoracic and sacral curves are called convex curves. This is not true, however, of a newborn baby's spine. It forms a continuous convex curve from top to bottom. Gradually, as the baby learns to hold up his head, a reverse or concave curve develops in his neck (cervical region). And later as the baby learns to stand, the lumbar region of his spine also becomes concave.

The normal curves of the spine serve important functions. They give it enough strength to support the weight of the rest of the body and the balance necessary for us to stand and walk on two feet instead of having to crawl on all fours. A curved structure has more strength than a straight one of the same size and materials. (The next time you pass a bridge look to see whether or not its supports form a curve.) Pretty clearly the spine needs to be a strong structure. It supports the head balanced on top of it, the ribs and internal organs suspended from it in front, and the hips and legs attached to it below.

Disease or poor posture often causes the lumbar curve to become abnormally exaggerated, a condition commonly called "swayback" or, technically, *lordosis.* Another abnormal curvature is *kyphosis,* known to most of us as "hunchback."

Thorax

Twelve pairs of ribs, the sternum (breast bone), and the thoracic vertebrae form the bony cage known as the thorax or chest. To find out what the terms "true ribs," "false ribs," "floating ribs," and "costal cartilage" mean, consult Table 2-2.

Upper and lower extremities

Well over a hundred bones form the framework of our arms, legs, shoulders, and hips. The upper arm and the upper leg (the thigh) contain only one bone each—the *humerus* in the upper arm and the *femur* in the thigh. Notice in Figure 2-2 how long these bones are. In fact, the femur is the longest bone in the body and the humerus is the second longest. Take another look at Figure 2-2 and note the names of the bones in the lower arm and lower leg. Feel the bone running up the thumb side of your lower arm. Is its name radius or ulna? The technical name for the large bone that forms a rather sharp edge along the front of your lower leg is the *tibia.* You may have called this your "shin bone." A slender, somewhat fragile bone, named the *fibula,* lies in the side portion of the lower leg.

The hands have more bones in them for their size than any other part of the body—

fourteen finger bones or *phalanges,* five *metacarpal* or palm-of-the-hand bones, and eight *carpal* or top-of-the-hand bones —a total of twenty-seven bones in each hand (Figure 2-4). This structural fact has great functional importance as we shall explain in the discussion of joints.

Toe bones have the same name, *phalanges,* as finger bones. There is the same number of toe bones as finger bones; a fact that might surprise you since toes are so much shorter than fingers. Foot bones comparable to the metacarpals and carpals of the hand have slightly different names. They are called *metatarsals* and *tarsals* in the foot (Figure 2-5). Just as each hand contains five metacarpal bones, each foot contains five metatarsal bones. But the foot has only seven tarsal bones in contrast to the hand's eight carpals. (A way that helps me remember which of these are foot bones and which are hand bones is to associate tarsal and toes together because they both begin with the letter "t.")

The *shoulder girdle* attaches the arms to the trunk and the *hip girdle* attaches the legs to the trunk. The two clavicles (collarbones) and the two scapulas (shoul-

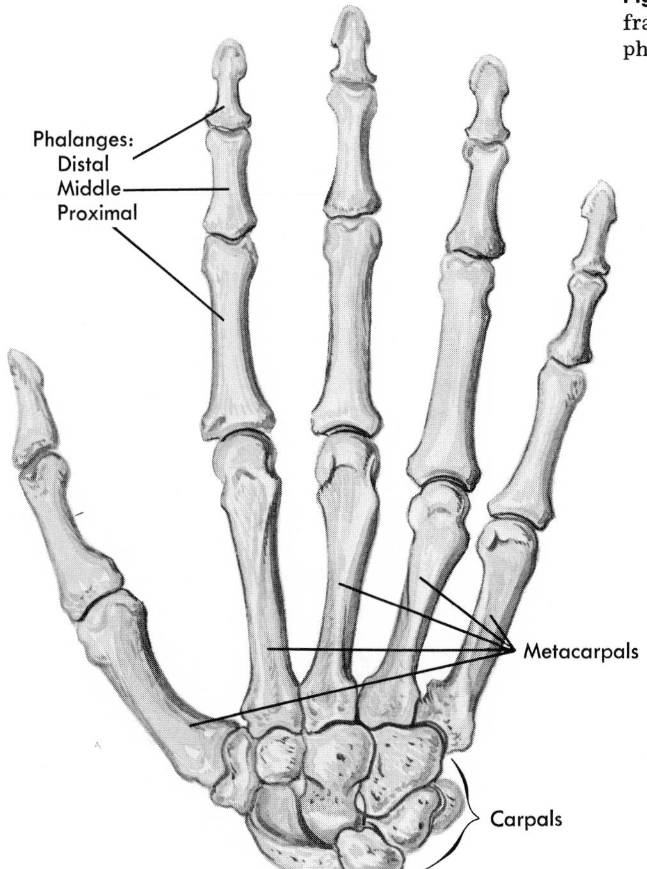

Fig. 2-4 Note that three phalanges form the framework of each finger. How many phalanges in the thumb?

der blades) form the shoulder girdle. The two large *innominate bones* (pelvic or hip bones) compose the hip girdle. In a baby's body each innominate bone consists of three bones—the ilium, ischium, and pubis. Later these bones grow together to become one bone in an adult.

Phalanges
Metatarsals
A
Tarsals

Differences between a man's and a woman's skeleton

A man's skeleton and a woman's skeleton differ in several ways. Were you to examine a male and a female skeleton placed side by side, you would probably notice first the difference in their sizes. Most male skeletons are larger than most female skeletons—a structural fact that seems to have no great functional importance. Structural differences between the male and female pelvic bones, however, do have functional importance. The female pelvis is made so that the body of a baby can be cradled in it as he grows, and when the time comes can pass through it in his birth journey. Although the individual male pelvic (innominate) bones are generally larger than the individual female pelvic bones, together the male pelvic bones form a narrower structure than do the female pelvic bones. A man's pelvis is shaped something like a funnel, but

Fig. 2-5 A, Compare the names and numbers of foot bones shown here with those of the hand bones shown in Fig. 2-4. **B,** The inner longitudinal arch of the foot, an arch from front to back that is formed by the arrangement of some of the metatarsal and tarsal bones.

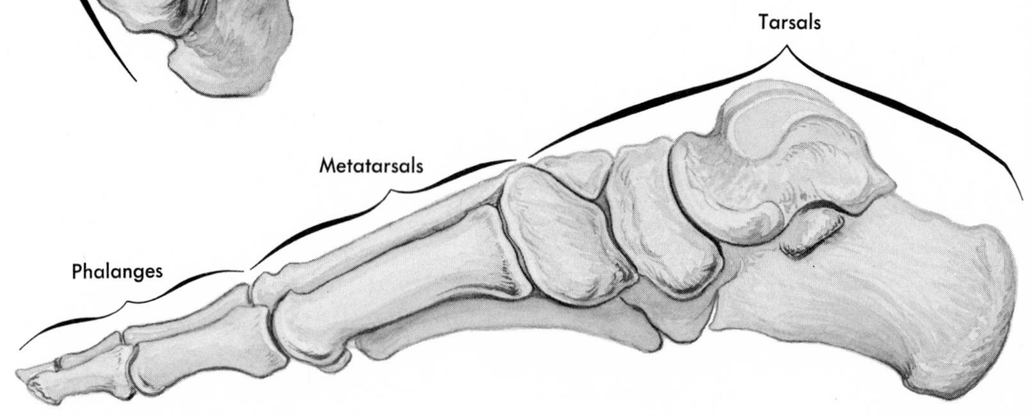

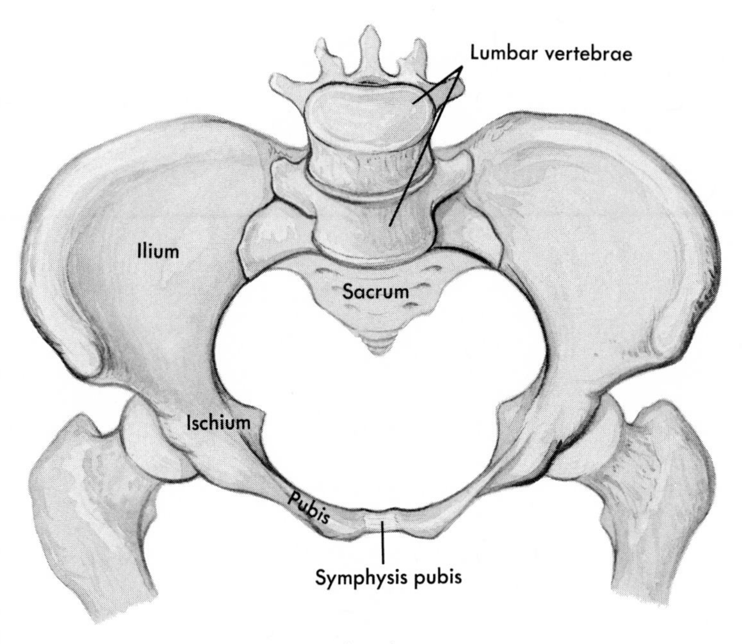

Fig. 2-6 Compare the shape of the female pelvis shown here with that of the male pelvis shown in Fig. 2-7.

Fig. 2-7 Note the narrower width of this male pelvis compared with the female pelvis shown in Fig. 2-6.

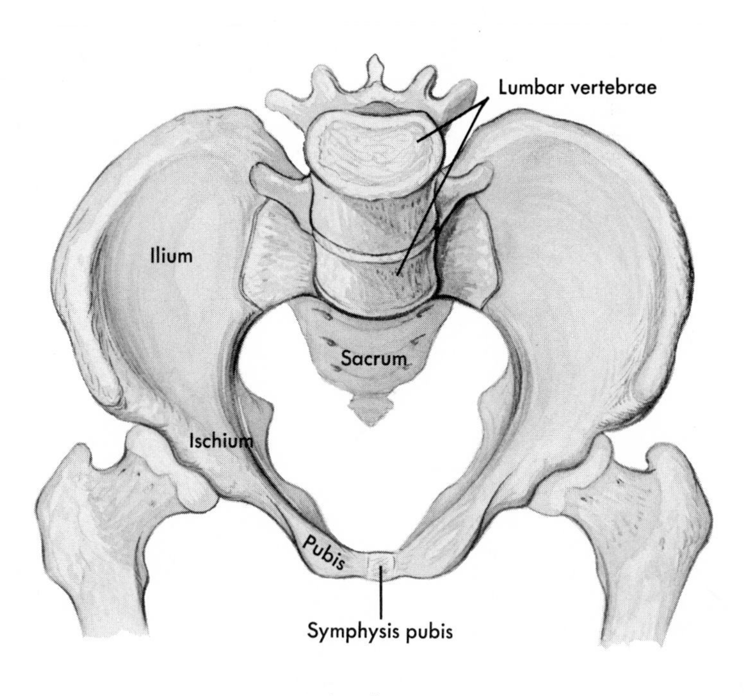

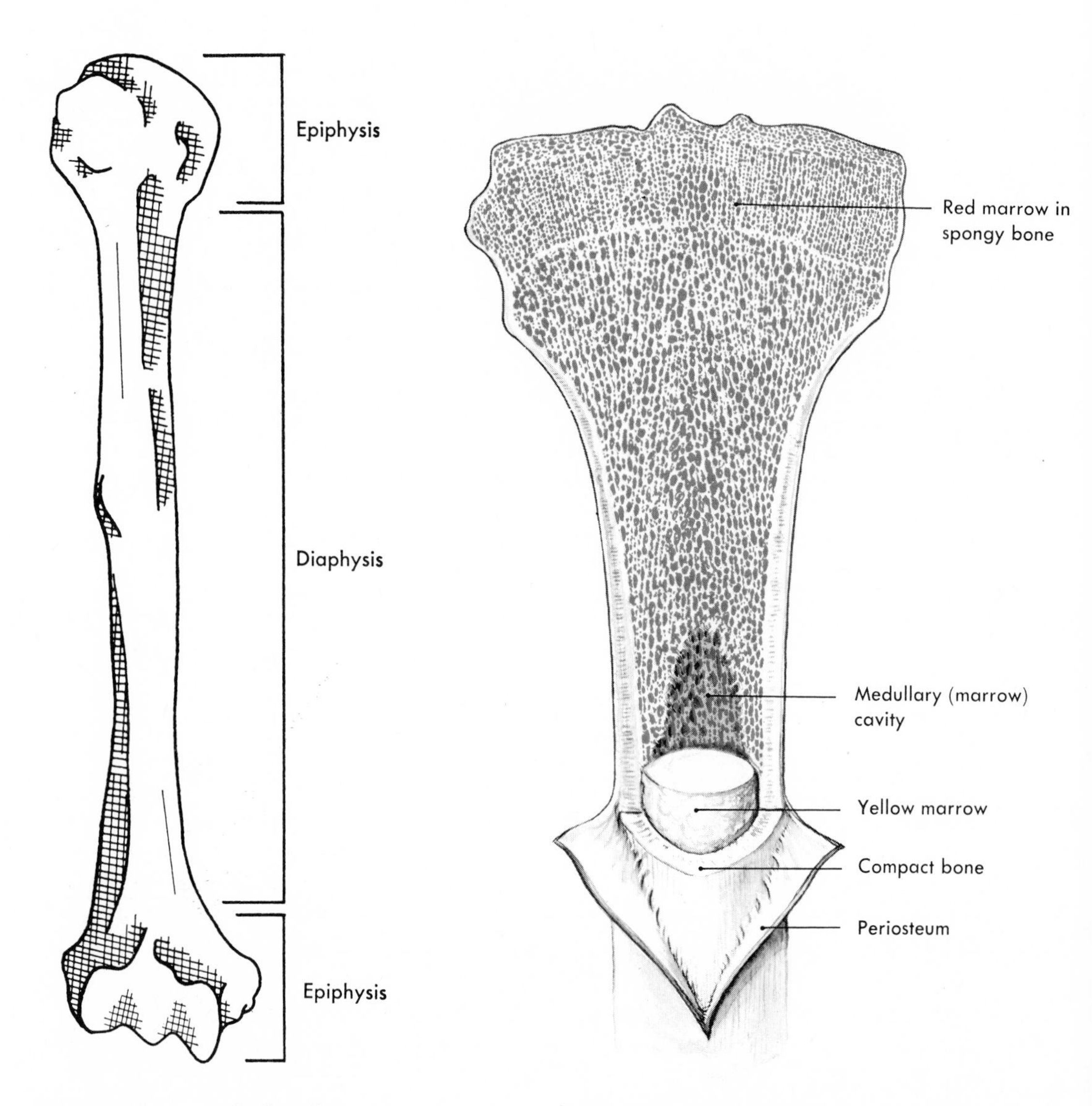

Fig. 2-8 Divisions of a long bone.

Fig. 2-9 Cutaway section of a long bone.

a woman's pelvis has a broader, shallower shape, more like a basin. (Incidentally, the word pelvis means basin.) Another difference is that the *pelvic inlet,* or brim, is normally much wider in the female than in the male. Figures 2-6 and 2-7 show this difference clearly. In these figures you can also see how much wider the angle is at the front of the female pelvis where the two pubic bones join than it is in the male.

Differences between a baby's and an adult's skeleton

Have you ever noticed that a newborn baby's head seems larger for the size

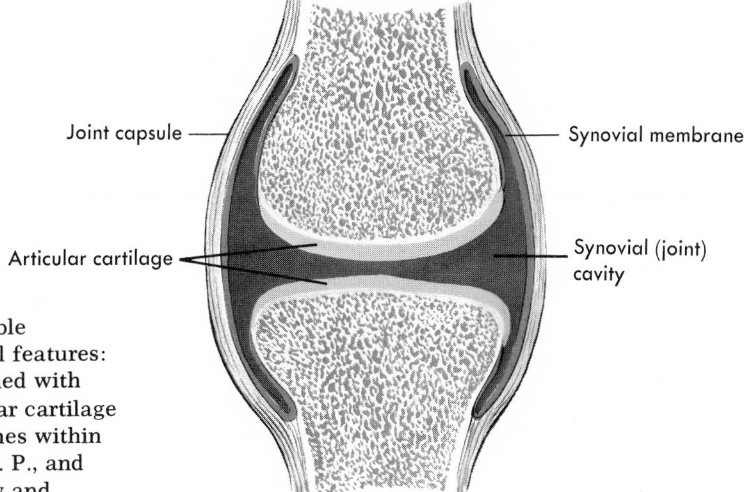

Fig. 2-10 Structure of a freely movable (diarthrotic) joint. Note these typical features: the joint capsule, the joint cavity lined with synovial membrane, and the articular cartilage covering the end surfaces of the bones within the joint capsule. (From Anthony, C. P., and Kolthoff, N. J.: Textbook of anatomy and physiology, ed. 8, St. Louis, 1971, The C. V. Mosby Co.)

of its body than an adult's head does? This is one of several differences between a baby's and an adult's skeleton. A newborn baby's head is about 1/4 as long as its whole body, whereas an adult's head is about 1/8 of its total height. A baby's chest is round; an adult's is oval. A baby's face appears small and its cranium large; an adult's face and cranium look nearly the same size.

At birth the skeleton is unfinished since some of the bones still consist partly of cartilage or fibrous tissue. Familiar examples of this are the soft spots (fontanels) in a baby's head. Another example is the cartilage between the shaft of a long bone and its ends. This is known as the *epiphyseal cartilage,* and it remains as long as bones are still growing. Epiphyseal cartilage can be seen in x-ray pictures—an important fact for a doctor who wants to know if a child is going to grow any more. No epiphyseal cartilage means no more growth in bone length.

Structure of long bones

Figures 2-8, and 2-9, and 2-10 will help you learn the names of the main parts of a long bone. Identify each of the following:

1. *Diaphysis,* or shaft—made of hard compact bone; is a hollowed-out cylinder filled with yellow bone marrow
2. *Epiphyses,* or the ends of the bone—red bone marrow fills in small spaces in the spongy bone composing the epiphyses
3. *Articular cartilage*—a thin layer of cartilage covering each epiphysis; serves much the same purpose that a small rubber cushion would if it were placed over the ends of bones where they come together to form a joint
4. *Periosteum*—a strong fibrous membrane that covers a long bone except at its joint surfaces where it is covered by articular cartilage
5. *Medullary cavity*—the hollow area inside the shaft of a bone; contains yellow bone marrow

Joints (articulations)

Every bone in the body, except one, connects to at least one other bone. In other words, every bone but one forms a joint with some other bone. (The exception is the hyoid bone in the neck, to which the tongue anchors.) Most of us never think much about our joints unless something goes wrong with them and they do not function properly. Then their tremendous importance becomes painfully clear. Joints perform two functions: they hold our bones together securely, and at the same time they make it possible for movement to occur between the bones—between most of them, that is. In a few places, though, joints are constructed so that no movement can take place. Most joints in the skull, for example, are immovable. The only movable joints in the skull are those between the lower jaw bone and the temporal bones. A few of the joints in the body allow only very slight movement, but by far the majority of them allow considerable movement—sometimes in many directions and sometimes in only one or two directions. Try, for example, to move your arm at your shoulder joint in as many directions as you can. Try to do the same thing at your elbow joint. Now examine the shape of the bones at each of these joints on a skeleton or in Figures 2-2 and 2-3. Do you see why you cannot move your arm at your elbow in nearly as many directions as you can at your shoulder?

Freely movable joints are all made alike in certain ways. All of them have a joint capsule, a joint cavity, and a layer of cartilage over the ends of two joining bones (Figure 2-10). The *joint capsule* is made of the body's strongest and toughest material—fibrous connective tissue—and is lined with smooth, slippery synovial membrane. The capsule fits over the ends of the two bones something like a sleeve. Because it attaches firmly to the shaft of each bone to form its covering (called the periosteum; "peri" means around and "osteum" means bone), the joint capsule holds the bones together securely but at the same time permits movement at the joint. The structure of the joint capsule, in other words, helps make possible the joint's function.

Ligaments (cords or bands made of the same strong fibrous connective tissue as the joint capsule) also grow out of the periosteum and lash the two bones together even more firmly.

The layer of *articular cartilage* over the joint ends of bones acts like a rubber heel on a shoe—it absorbs jolts. The *synovial membrane* secretes a lubricating fluid (synovial fluid) that allows easier movement with less friction.

Because of the structure of freely movable joints, we can move them in a variety of ways. These will be discussed with muscle action in the next chapter.

Diseases of bones

If someone asked you to name a bone disease perhaps you would quickly answer "rickets." Years ago many more children developed this disease than do today. Rickets occurs when a child's blood contains too little "vitamin" D.* With a deficiency

*Scientists now tell us that vitamin D is really not a vitamin at all but a hormone released from the skin when the ultraviolet rays of the sunlight play upon the skin. They have named this substance calciferol; its nickname is the "bone-calcifying hormone." See Loomis, W. F.: Rickets, Sci. Amer. 223:77-91, 1970.

of vitamin D too little calcium deposits in bones and they fail to harden, that is, calcify, normally. One unfortunate and all too evident result is that the child's leg bones gradually bend from bearing the weight of his body and he becomes bow-legged.

Bone diseases occur even more often in elderly people than in children. For instance old people's bones usually become more porous than normal. This change may make the bones so weak that they break under the body's weight—a fact that helps to explain many of the broken hips in old people. Exactly what causes *osteoporosis* (condition in which bones are more porous than normal) no one yet knows for sure, but too little exercise and too small amounts of sex hormones presumably have a good deal to do with it. Consider these facts as evidence: from the menopause on, women characteristically have an estrogen deficiency and usually exercise sparingly. Most of them also have some degree of osteoporosis. In short, menopause, deficient estrogens, deficient exercise, and osteoporosis more often than not go together. Osteoporosis is also common in older men.

Diseases of joints

Almost everyone sooner or later suffers from some kind of arthritis in some part of his body. *Arthritis*, or joint inflammation, cripples hundreds each year. Medical science has not yet established its exact causes nor has it learned how to defeat this powerful and prevalent enemy. Advancing years take their toll on the joints by causing degenerative changes *(osteoarthritis)* in them. This is especially true for the weight-bearing joints and for the joints of overweight individuals. The articular cartilages (Figure 2-10) undergo degenerative changes in osteoarthritis, and excess bone grows along the joint edges of bones. These accumulations of bone are called "spurs" and "marginal lipping." The presence of osteoarthritis undoubtedly accounts for many of the complaints made by older people, such as, "My joints are getting so stiff I even have trouble getting up out of a chair" or "I'm not nearly as spry as I used to be; I can't move around so easily."

Among other kinds of frequently occurring arthritis are *rheumatoid arthritis* and *gout*.

outline summary

FUNCTIONS

1 Supports and gives shape to body
2 Protects internal organs
3 Helps make movements possible

GENERAL PLAN

Skeleton composed of following two main divisions and their subdivisions:

1 Axial skeleton
  a Skull
  b Spine
  c Thorax
  d Ear bones
  e Hyoid bone
2 Appendicular skeleton
  a Upper extremities, including shoulder girdle
  b Lower extremities, including hip girdle

LOCATION AND DESCRIPTION OF BONES

See Figures 2-1, 2-2, 2-3, and Table 2-2

Differences between a man's and a woman's skeleton

1 In size—male skeleton generally larger
2 In shape of pelvis—male pelvis deep and narrow, female pelvis broad and shallow
3 In size of pelvic inlet—female pelvic inlet generally wider, normally large enough for baby's head to pass through
4 In pubic angle—angle between pubic bones of female generally wider

Differences between a baby's and an adult's skeleton

1 Baby's head proportionately longer—about 1/4 total length of body; length of adult head about 1/8 total height of body

2 Baby's face smaller than cranium; adult's face and cranium about the same size
3 Baby's chest is round, adult's is oval
4 Parts of baby's skeleton consist of cartilage or fibrous tissue such as "soft spots" in skull

STRUCTURE OF LONG BONES

See Figures 2-8, 2-9

JOINTS (ARTICULATIONS)

1 Functions—hold bones together securely; make possible movements
2 Structures of freely movable joints
  a Joint capsule and ligaments hold joining bones together but permit movement at joint
  b Articular cartilage—covers joint ends of bones and absorbs jolts
  c Synovial membrane—lines joint capsule and secretes lubricating fluid
  d Joint cavity—space between joint ends of bones

DISEASES OF BONES

1 Rickets—too little calcium deposits in bones because of deficiency of calciferol (vitamin D, the "bone-calcifying hormone") in blood
2 Osteoporosis—bones become more porous than normal presumably because of too little exercise and too small amounts of sex hormones

DISEASES OF JOINTS

1 Arthritis—joint inflammation; causes not yet established
2 Osteoarthritis—articular cartilages of joint undergo degenerative changes and excess bone grows along the joint edges of bone; called "spurs" and "marginal lipping"
3 Other types of arthritis—rheumatoid arthritis and gout, for example

new words

arthritis
articulation
appendicular skeleton
axial skeleton
cervical
diaphysis
epiphyses
kyphosis
ligament
lordosis
lumbar
osteoarthritis
osteoporosis
periosteum
rickets
sinus
sinusitis
synovial membrane
thorax

review questions

1 What functions does the skeletal system perform for the body?
2 Give the anatomical name of the following:
   bone on little finger side of lower arm
   collarbone
   finger bones
   forehead bone
   lower jaw bone
   shoulder blade
   thigh bone
   toe bones
   upper arm bone
3 Name the small bones in the middle ear.
4 What functions do joints perform?
5 Describe the structure of a freely movable joint.
6 Locate and briefly describe each of the following bones:
   clavicle
   ethmoid
   femur
   frontal
   humerus
   ilium
   innominate
   mandible
   metacarpals
   parietal
   patella
   phalanges
   radius
   scapula
   sphenoid
   sternum
   tarsals
   tibia
   turbinates
   zygomatic
   ulna
7 Describe one marked difference between a male and a female adult skeleton.
8 Describe some differences between an infant and an adult skeleton.
9 Is it possible to tell whether a child is going to grow any taller? If so, how can a doctor tell this?
10 What structural changes frequently occur in older people's joints?

3

The muscular system

When we speak of the muscular system, we mean the more than 500 muscles by which we move in many ways, varying in complexity from blinking an eye or smiling to climbing a mountain or ski-jumping. Not many of our body structures can claim as great an importance for happy useful living as can our voluntary muscles, and only a few can boast of greater importance for life itself. A great deal is known about muscles—enough, in fact, to fill several books the size of this one. So in this chapter we shall try to present only information that practical nurses and other beginning students of the body will find useful to know. The plan is this—to investigate first the different types of muscle tissue, then to note some general facts about the structure and function of skeletal muscles, next to present some specific facts about certain key muscles and about posture, and finally to consider some muscle disorders.

Muscle tissue

If you weigh 120 pounds, about 50 pounds of your weight comes from your muscles, the "red meat" attached to your bones. Under the microscope these muscles appear as bundles of fine threads with many crosswise stripes. Each fine thread is a muscle cell or, as it is usually called, a *muscle fiber*. This type of muscle tissue has three names: *striated muscle*—because of its cross stripes or stria, *skeletal muscle*—because it attaches to bone, and *voluntary muscle*—because its contractions can be controlled voluntarily.

Beside skeletal muscle, the body also contains two other kinds of muscle tissue, and each of these also has more than one name. One kind is called *branching muscle*—because its cells seem to branch into each other—or *cardiac muscle*—because it composes the bulk of the heart. The third kind of muscle tissue consists of slender, tapered cells that have no cross stripes.

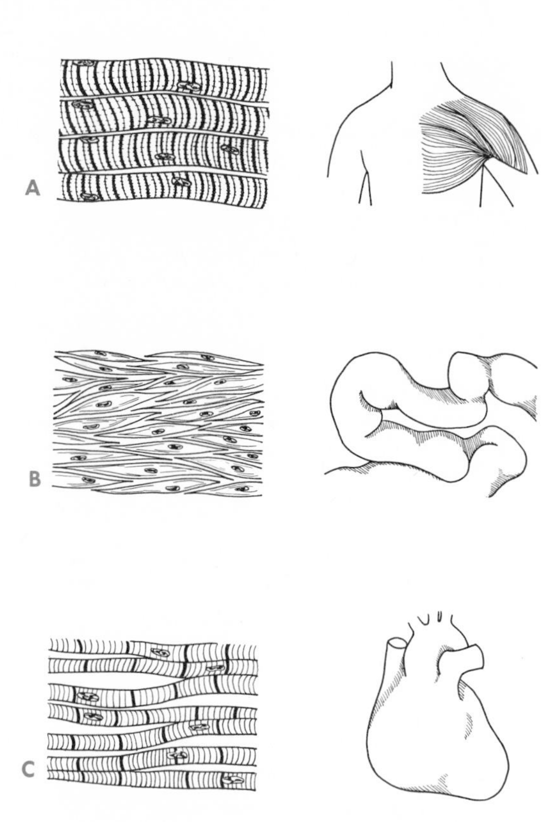

Fig. 3-1 Muscle tissues of the human body. **A,** Striated, or skeletal, muscle tissue; left, a microscopic view showing cross striations and multinuclei per cell; right, a macroscopic view of skeletal muscle organs. **B,** Nonstriated, or smooth, muscle tissue; left, a microscopic view; right, a loop of intestine, one of many internal organs whose walls contain smooth muscle. **C,** Branching, or cardiac, muscle tissue; left, a microscopic view showing cross striations and branching cells; right, the heart, the only organ made of cardiac muscle tissue.

Nonstriated muscle and *smooth muscle* are the names given to this tissue. It is also called *involuntary muscle*—for the obvious reason that most people cannot voluntarily control the contraction of this type of muscle. Some rather startling exceptions to this rule do exist, however. For instance, there are a few people who can do such things as make the hair on their arms stand on end or their pupils dilate at will. Contraction of smooth (involuntary) muscle brings about both of these actions. Smooth muscle forms an important part of blood vessel walls and of many internal organs (viscera), so another name for it is *visceral muscle* (Figure 3-1).

Muscle cells specialize in the function of contraction, or shortening. Every movement we make is produced by the contraction of muscle cells.

Skeletal muscles (organs)

A skeletal muscle is an organ composed mainly of striated muscle cells and connective tissue. Most skeletal muscles attach to two bones that have a movable joint between them. In other words, most muscles extend from one bone across a joint to another bone. Also, one of the two bones moves less easily than the other. The muscle's attachment to this more stationary bone is called its *origin*. Its attachment to the more movable bone is called the muscle's *insertion*. The rest of the muscle (all of it except its two ends) is called the *body* of the muscle. *Tendons* anchor muscles firmly to bones. Made of dense fibrous connective tissue in the shape of heavy cords, tendons have great strength. They do not tear or pull away from bone easily. Yet any emergency-room nurse or doctor sees many tendon injuries—both severed tendons and tendons torn loose from bones.

Small sacs called *bursae* lie between some tendons and the bones beneath them. Bursae are made of connective tissue lined with synovial membrane and contain a small amount of synovial fluid, a slippery, lubricating substance. Like a small, flexible cushion, a bursa makes it easier for a tendon to slide over a bone when the tendon's muscle shortens. *Tendon sheaths* enclose certain tendons. Because these tube-shaped structures are lined with synovial membrane that is moistened with synovial fluid, they, like the bursae, facilitate movement. Both bursae and tendon sheaths can become inflamed. Inflammation of a bursa – a relatively common ailment, particularly in older people – is called *bursitis*. The term that means inflammation of a tendon sheath is *tenosynovitis*.

Muscles move bones by pulling on them. Because the length of a skeletal muscle becomes shorter as its fibers contract, the bones the muscle attaches to move nearer together. As a rule, only the insertion bone moves. The shortening of the muscle pulls the insertion bone toward the origin bone. The origin bone stays put, holding firm while the insertion bone moves toward it. One tremendously important function of skeletal muscle contractions, therefore, is to produce movements. Remember this rule: a muscle's insertion bone moves toward its origin bone. It can help you understand muscle actions.

Another important fact to remember about the functioning of skeletal muscles is this: they generally work in teams, not singly. Several muscles contract at the same time to produce almost any movement you can think of. Of the muscles contracting simultaneously, the one mainly responsible for producing the movement is called the *prime mover* for that movement. The other muscles that help produce the movement are called *synergists*. For example, the prime mover for extension of the lower leg is the rectus femoris muscle of the thigh. The synergists that help the rectus femoris bring about this movement are various other muscles located on the front of the thigh.

From what we have said so far, you may have formed the idea that only your skeletal muscles have to function for you to move. This is not so at all. Many other structures must function with them. For instance, as you have already observed, most muscles bring about movements by pulling on bones across movable joints. But before a skeletal muscle can contract and pull on a bone to move it, the muscle must first be stimulated by nerve impulses. It must also be supplied with oxygen and food; otherwise it soon cannot contract because of a lack of energy. Energy for all kinds of work done by all kinds of body cells, you will recall, comes from the cells' catabolism of foods. Catabolism uses oxygen, and in order for cells to receive their vital oxygen supply, the respiratory system must move oxygen from the air into the blood. The circulatory system must then deliver the oxygen to the body cells. In short, the respiratory, circulatory, nervous, muscular, and skeletal systems all play essential parts in producing normal movements. This fact has great practical importance. For example, a person might have perfectly normal muscles and still not be able to move normally. He might have a nervous system disorder that shuts off impulses to certain skeletal muscles and thereby paralyzes them. Poliomyelitis is one great enemy that acts in this way, but so do some other conditions – a hemorrhage in the brain, a brain tumor, or a spinal cord injury, to mention a few. Skeletal system disorders, arthritis especially, can have disabling effects on movement. Muscle functioning, then, depends upon the functioning of many other parts of the body. This fact illustrates a principle repeated often in this book. It can be simply stated: no part of the body lives by or for itself alone. Each part depends upon all other parts for its healthy survival. Each part contributes something to the healthy survival of all other parts.

Contraction of a skeletal muscle does not

always produce movement. Sometimes it increases the tension within a muscle but does not decrease the length of the muscle. When the muscle does not shorten, no movement results. One type of contraction that produces no movement is known as *isometric contraction.* The word isometric comes from Greek words that mean "equal measure." In other words, a muscle's length during an isometric contraction and during relaxation are about equal. Although muscles do not shorten (and therefore do not produce movement) during isometric contractions, the tension within them increases. Because of this, repeated isometric contractions tend to make muscles grow larger and stronger. Hence the popularizing in recent years of isometric exercises as great muscle builders. The type of muscle contraction that produces movement is *isotonic contraction.* Try this simple experiment to see for yourself the difference between isometric and isotonic contractions. Watch the front of your upper arm as you bend your arm at the elbow and move your hand up toward your shoulder. As you make this movement, can you see the muscle in your upper arm become shorter and more bulging? Now place one hand on the undersurface of a heavy table or desk top, and, still watching your upper arm, push up on the table top with all your strength. Did your upper arm muscle (biceps brachii is its name) shorten this time as it contracted? Did it produce movement? What kind of contraction did you perform this time—isometric or isotonic? Answer question 7 on p. 49.

Muscle tone, or tonic contraction, is another type of skeletal muscle contraction that produces no movement. Because relatively few of a muscle's fibers shorten at one time in a tonic contraction, the muscle as a whole does not shorten. Consequently, tonic contractions do not move any body parts. What they do is hold them in position. Or, using synonyms for the words tonic contractions and position, muscle tone maintains *posture.* Posture and movements are the two great functions of skeletal muscles. Both contribute greatly to our healthy survival. Good posture means that body parts are held in the positions that favor best function. It means positions that balance the distribution of weight and that, therefore, put the least strain on muscles, tendons, ligaments, and bones. To have good posture in a standing position, for example, you must stand with your head and chest held high, your chin, abdomen, and buttocks pulled in, and your knees bent slightly.

To judge for yourself how important good posture is, consider some of the effects of poor posture. Besides detracting from appearance, poor posture makes a person tire more quickly. It puts an abnormal pull on ligaments, joints, and bones and therefore often leads to deformities. Poor posture crowds the heart, making it harder for it to contract. Poor posture crowds the lungs, decreasing their breathing capacity.

Skeletal muscles maintain posture by counteracting the pull of gravity. Gravity tends to pull the head and trunk down and forward, but certain back and neck muscles pull just hard enough in the opposite direction to overcome the force of gravity and hold the head and trunk erect. Without continuous muscular pull on the leg bones our knees would collapse and we would not be able to stand up.

Movements, skeletal muscles' other function, are even more important for health than good posture is. Most of us believe that "exercise is good for us" even if we have no idea what or how many specific benefits can come from it. The benefits of exercise are almost legion. Professional journals carry articles about exercising, and so do many popular magazines and books.* Some of the good consequences of regular, properly practiced exercise are greatly improved muscle tone, better posture, more efficient heart and lung functioning, less fatigue, and, last but not least, looking and feeling better.

Sick people just naturally move about less than well people. Medically speaking, they become less mobile or even immobile. Too little mobility inevitably has bad effects on virtually all body structures and functions. These effects often threaten survival and sometimes lead to death. Good nursing care, however, can do much to hold off or lessen the damages of diminished activity. A nurse can teach her patients what and how much exercising they can do—even if bedfast†—and she can help and encourage them as needed. Later chapters will call your attention to the many harmful effects of immobility.

*Cooper, K. H.: How to feel fit at any age, Reader's Digest, 79-87, March 1968; Bowerman, W. J., and Harris, W. E.: Jogging, New York, 1967, Grosset & Dunlap.

†Kelly, M. M.: Exercises for bedfast patients, Amer. J. Nurs. 66(4):2209-2212, 1966.

Types of movements produced by skeletal muscle contractions

Muscles can move some body parts in several directions and others in only two directions. As mentioned on p. 34, this depends largely on the shapes of the bones at the freely movable joint. Some of the movements we make most often are flexion, extension, abduction, and adduction.

Suppose that you bend one of your arms at the elbow or bend a leg at the knee, or suppose that you bend over forward at the waist or bend your head forward in prayer. With each of these movements you will have flexed some part of the body—the lower arm, the lower leg, the trunk, and the head, respectively. Most flexions are movements that we commonly describe as bending. The only exception to this rule that I can think of is flexion of the upper

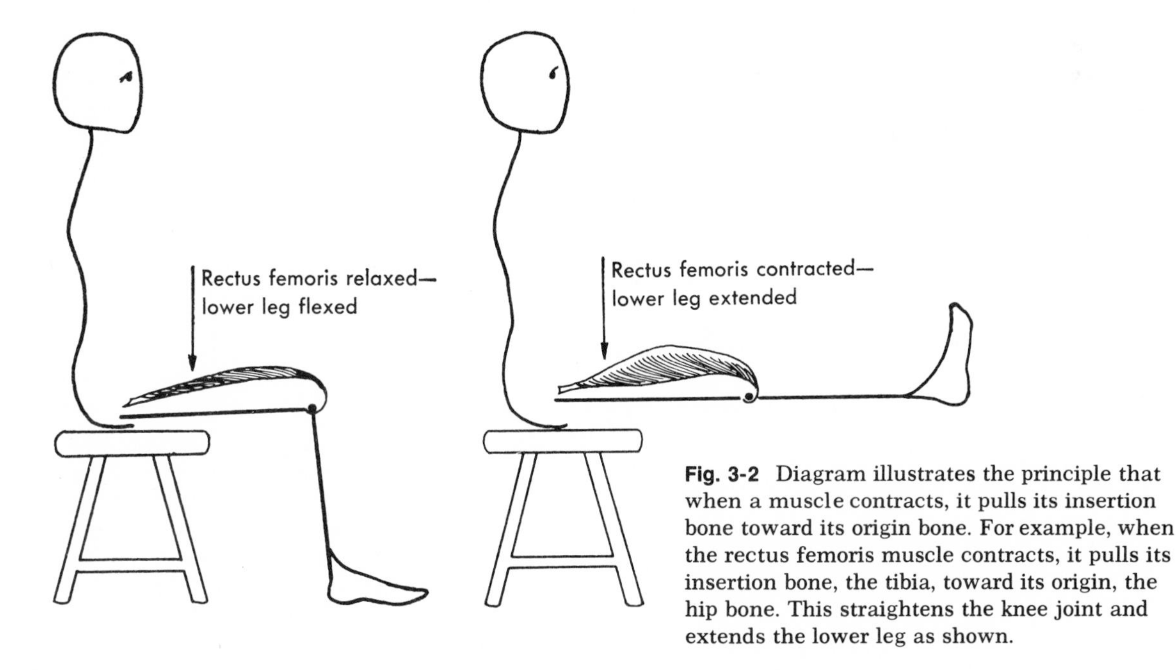

Fig. 3-2 Diagram illustrates the principle that when a muscle contracts, it pulls its insertion bone toward its origin bone. For example, when the rectus femoris muscle contracts, it pulls its insertion bone, the tibia, toward its origin, the hip bone. This straightens the knee joint and extends the lower leg as shown.

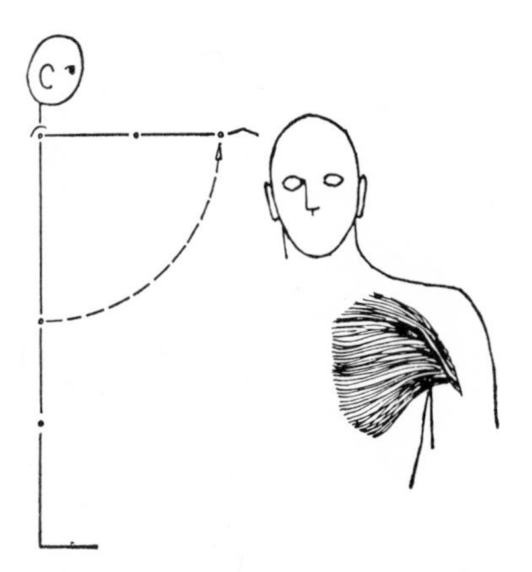

Fig. 3-3 When the pectoralis major muscle (shown on the figure at the right) contracts, it flexes the upper arm at the shoulder joint (figure at the left). What bone must the pectoralis major insert in if it moves the upper arm?

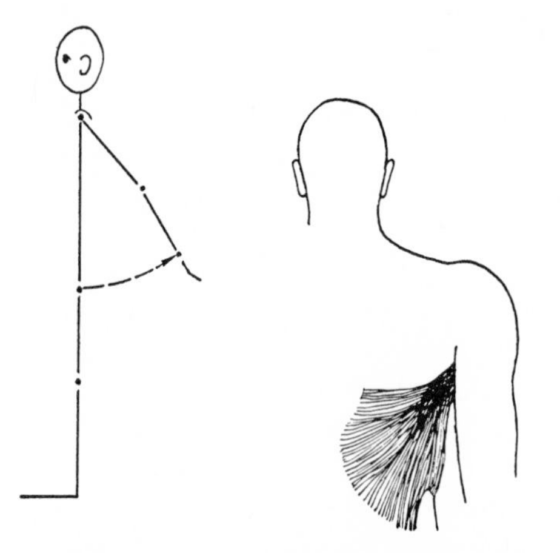

Fig. 3-4 When the latissimus dorsi muscle (shown on the figure at the right) contracts, it extends the upper arm at the shoulder joint (figure at the left). What bone must the latissimus dorsi insert in if it moves the upper arm?

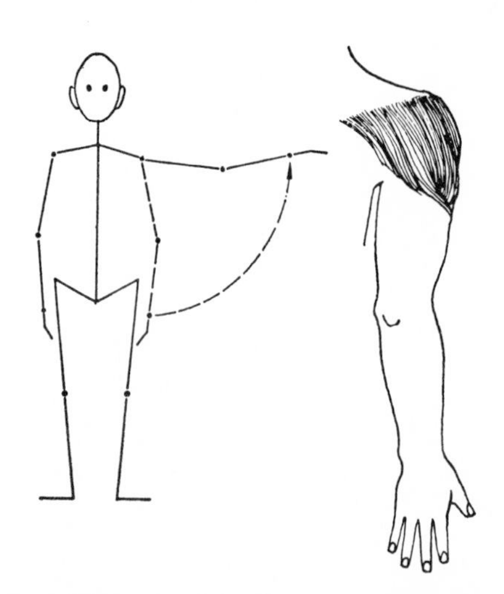

Fig. 3-5 When the deltoid muscle (shown on the figure at the right) contracts, it abducts the upper arm at the shoulder joint (figure at the left). In what bone must the deltoid insert to produce this movement?

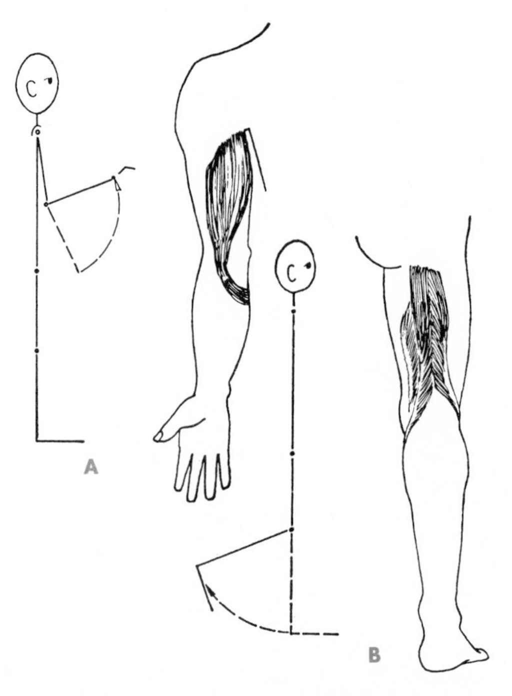

Fig. 3-6 Flexion of the lower arm and lower leg. **A,** When the biceps brachii muscle (shown at the right) contracts, it flexes the lower arm at the elbow joint (shown at the left). **B,** When the hamstring muscles (shown at the right) contract, they flex the lower leg at the knee joint (shown at the left).

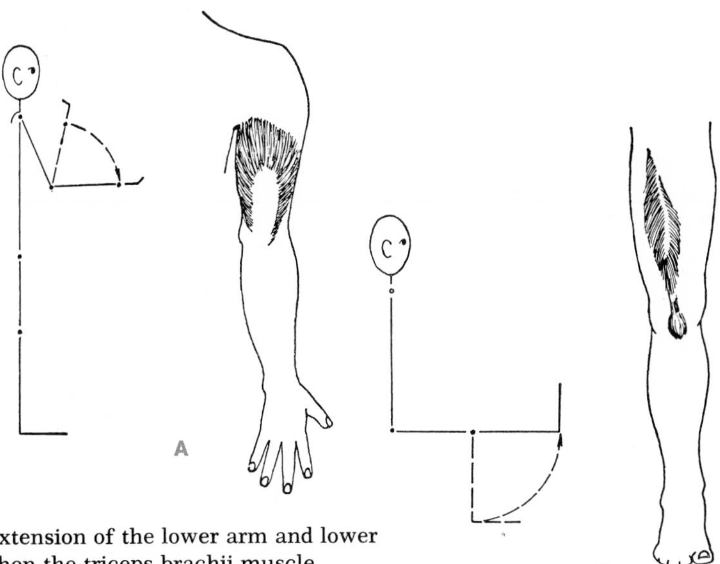

Fig. 3-7 Extension of the lower arm and lower leg. **A,** When the triceps brachii muscle (shown at the right) contracts, it extends the lower arm at the elbow joint (shown at the left). **B,** When the rectus femoris muscle (shown at the right) contracts, it extends the lower leg at the knee joint (shown at the left).

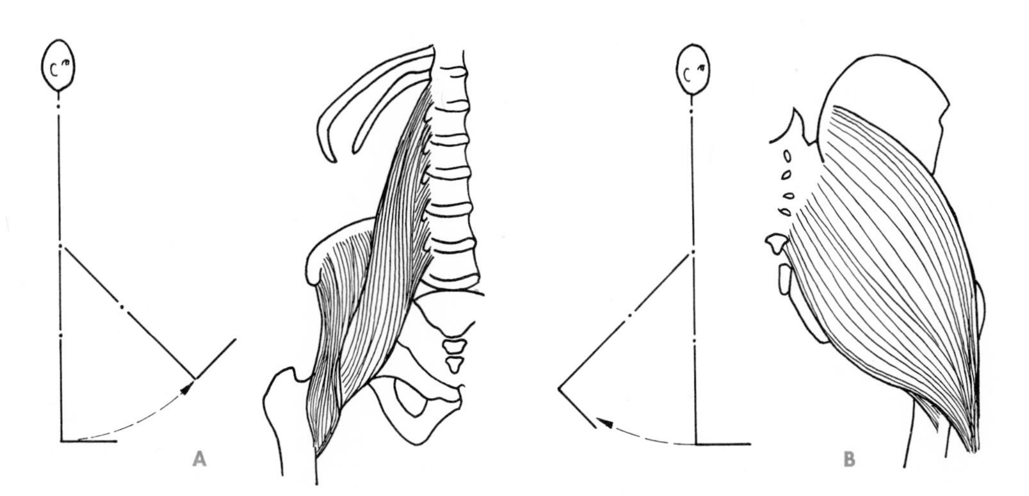

Fig. 3-8 Flexion and extension of the thigh. **A,** When the iliopsoas muscle (shown at the right) contracts and the femur serves as its insertion, it flexes the thigh at the hip joint (shown at the left). **B,** When the gluteus maximus muscle (shown at the right) contracts, it extends the thigh at the hip joint (shown at the left).

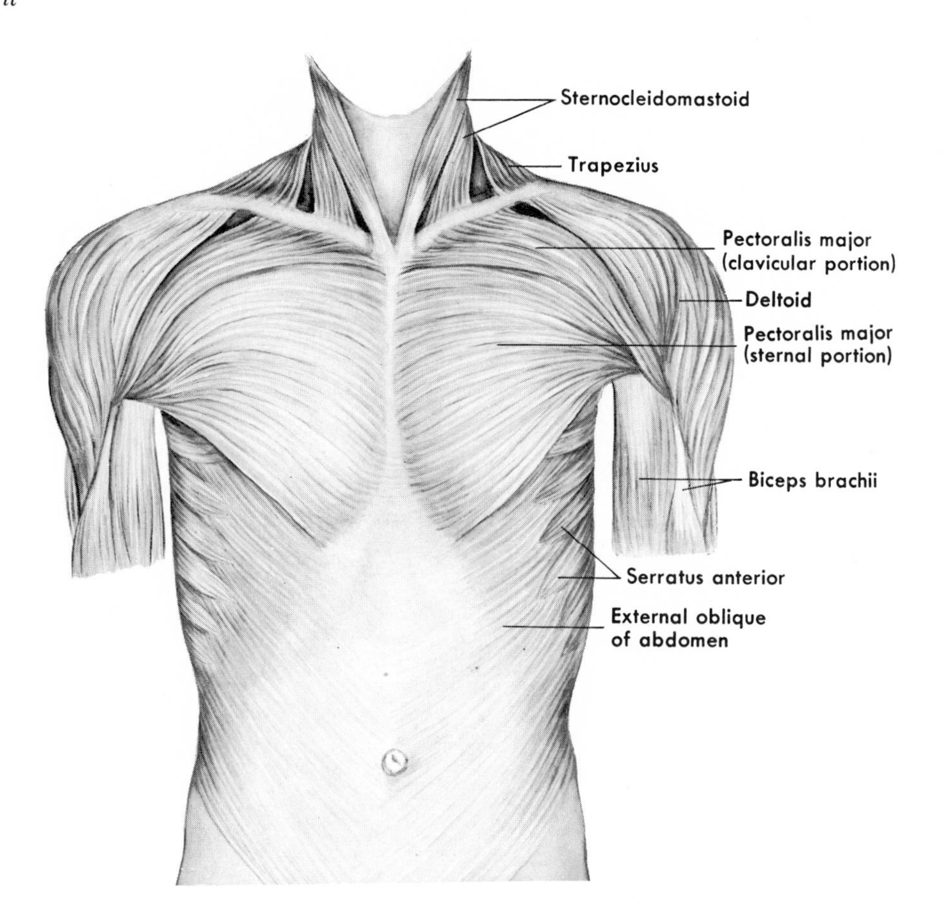

Fig. 3-9 Outer layer of the muscles of the front surface of the trunk. (From Francis, C. C: Introduction to human anatomy, ed. 5, St. Louis, 1968, The C. V. Mosby Co.)

arm. Flexing the upper arm means moving it forward from the chest, as shown in Figure 3-3. One definition for *flexion* is a movement that makes the angle between two bones at their joint smaller than it was at the beginning of the movement.

Extensions are opposite, or antagonistic, actions to flexions. Thus, extensions are movements that make joint angles larger rather than smaller. Extensions are straightening or stretching movements rather than bending movements. Extending the lower arm, for example, is straightening it out at the elbow joint, as shown in Figure 3-7, *A*. Extending the upper arm is stretching it out and back from the chest as shown in Figure 3-4. What movement would you make to extend your lower leg? to extend your thigh or upper leg? Look at Figures 3-7, *B*, and 3-8, *B*, to check your answers. If you straighten your back to stand tall or stretch backwards from your waist, are you flexing or extending your trunk?

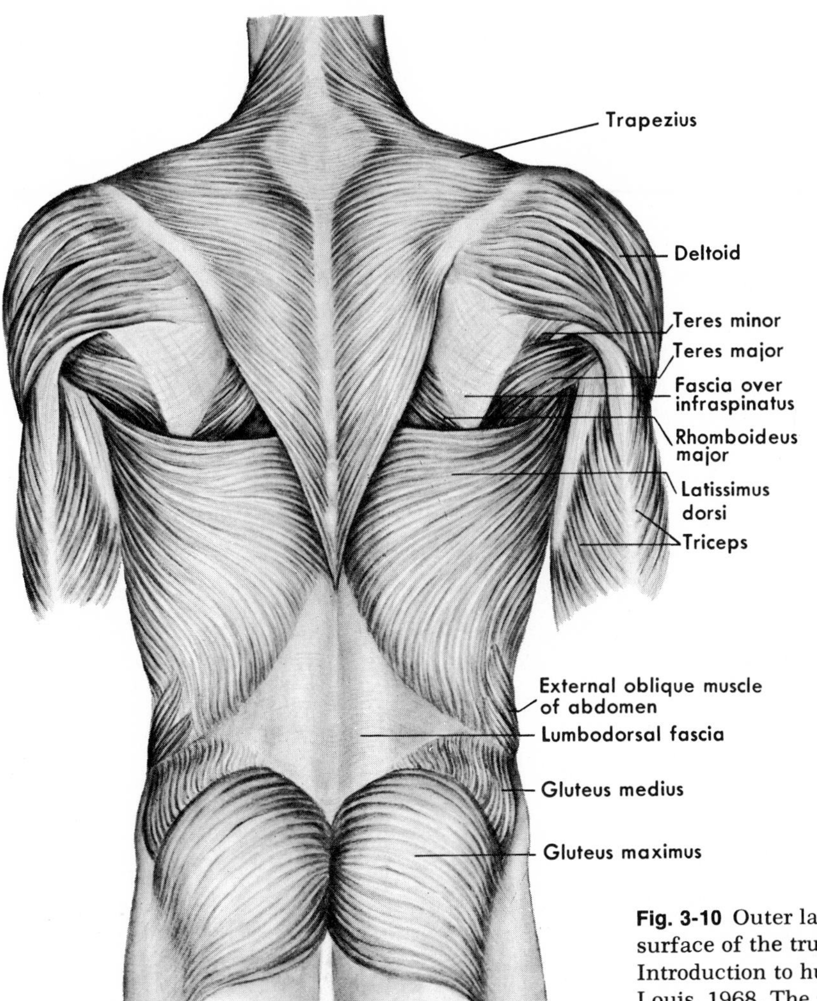

Fig. 3-10 Outer layer of the muscles of the back surface of the trunk. (From Francis, C. C: Introduction to human anatomy, ed. 5, St. Louis, 1968, The C. V. Mosby Co.)

Abduction means moving a part away from the midline of the body, such as moving the arms out to the sides.

Adduction means moving a part toward the midline, such as bringing the arms down to the sides.

Figures 3-9 to 3-12 show the outer layer of muscles known as the *superficial muscles*. *Deep muscles* lie under most of these.

Specific facts about certain key muscles

Flexors (that is, muscles that flex different parts of the body) produce many of the movements used in walking, sitting, swimming, typing, and a host of other activities. Extensors also function in these activities

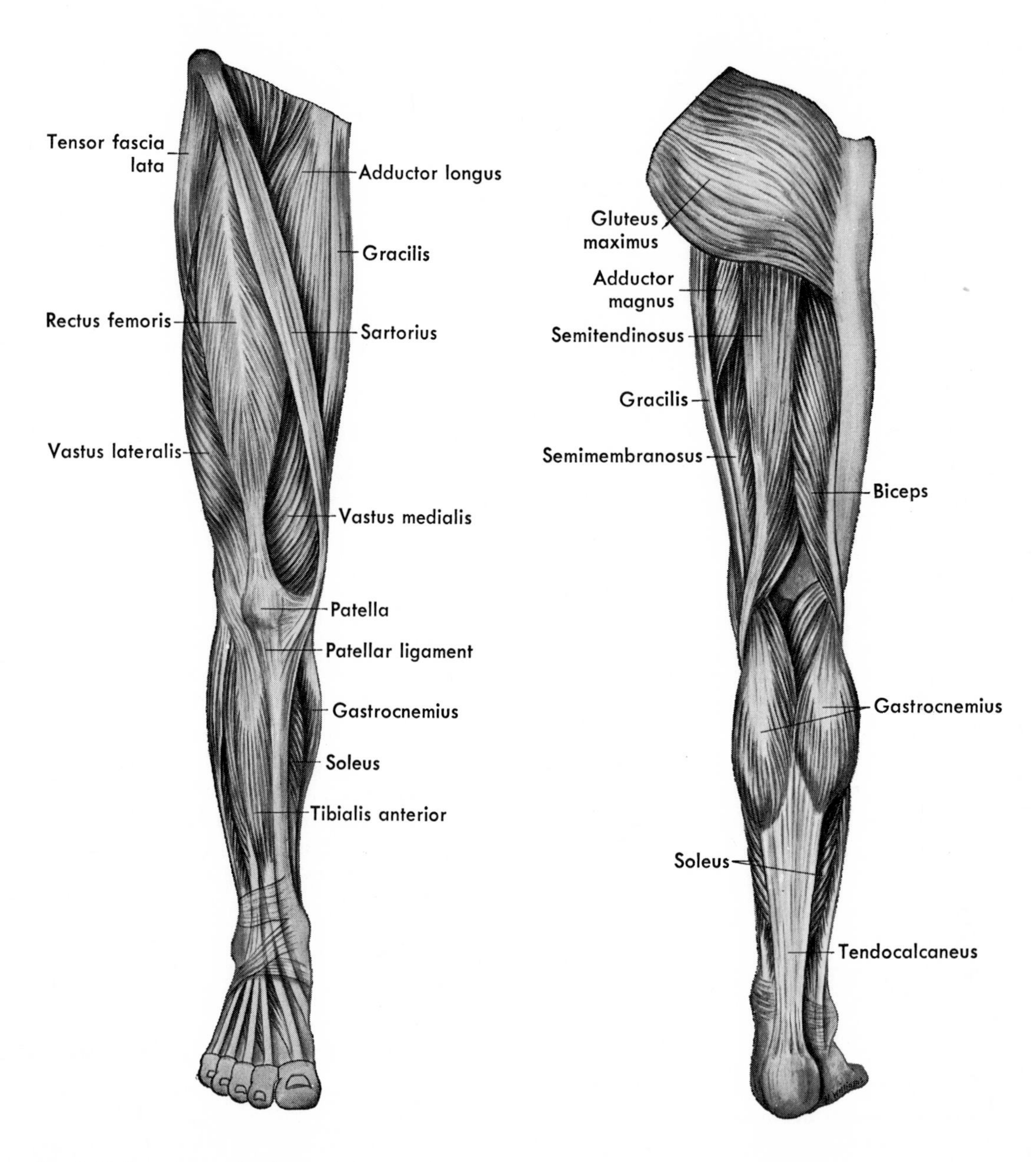

Fig. 3-11 Muscles of the front surface of the leg.

Fig. 3-12 Muscles of the back surface of the leg.

Table 3-1. Muscles grouped according to function

Part moved	*Flexors*	*Extensors*	*Abductors*	*Adductors*
Upper arm	Pectoralis major	Latissimus dorsi	Deltoid	Pectoralis major and latissimus dorsi contracting together
Lower arm	Biceps brachii	Triceps brachii	None	None
Thigh	Iliopsoas	Gluteus maximus	Gluteus medius and minimus	Adductor group
Lower leg	Hamstrings	Quadriceps femoris group	None	None
Foot	Tibialis anterior	Gastrocnemius soleus	Peroneus longus	Tibialis anterior
Trunk	Iliopsoas and rectus femoris	Erector spinae (sacrospinalis)	Psoas major and quadratus lumborum	Psoas major and quadratus lumborum

but perhaps play their most important role in maintaining upright posture. Study Table 3-1 and Figures 3-9 to 3-12 to learn the names of some of the prime flexors, extensors, abductors, and adductors of the body. Consult Table 3-2 to learn their origins and insertions. Keep in mind that muscles move the bone that they have their insertion on. You might use this information in this way—when you learn that a certain muscle is a flexor of the upper arm, you will know immediately that it must insert on what bone? If a muscle is listed as an extensor of the lower leg, what bone or bones must it insert on? Answer review questions 9 and 10 on p. 49.

Muscle disorders

Perhaps the best known disease of muscles is *muscular dystrophy.* It is a long-lasting disease whose main characteristics are progressive wasting and weakening of the muscles.

A person with *paralysis* cannot contract his muscles when he wants to; his skeletal muscles do not respond to his will. This is not because there is anything wrong with his muscles but because there is disease or injury of his brain or spinal cord or nerves so that they cannot activate his muscles.

Muscle atrophy is muscle shrinkage, a decrease in muscle size caused by disuse. Many conditions can cause atrophy, such as having a leg in a cast, being a bed patient for a long time, or paralysis.

Muscle hypertrophy is the opposite of atrophy; that is, it is an increase in size resulting from increased use. Skeletal muscles may hypertrophy as a result of exercise. The heart frequently hypertrophies from overwork.

Table 3-2. Muscle functions, origins, and insertions

Muscle	*Function*	*Insertion*	*Origin*
Pectoralis major	Flexes *upper arm* Helps adduct *upper arm*	Humerus	Sternum Clavicle Upper rib cartilages
Latissimus dorsi	Extends *upper arm* Helps adduct *upper arm*	Humerus	Vertebrae Ilium
Deltoid	Abducts *upper arm*	Humerus	Clavicle Scapula
Biceps brachii	Flexes *lower arm*	Radius	Scapula
Triceps brachii	Extends *lower arm*	Ulna	Scapula Humerus
Iliopsoas	Flexes *trunk*	Ilium Vertebrae	Femur
Iliopsoas	Flexes *thigh*	Femur	Ilium Vertebrae
Gluteus maximus	Extends *thigh*	Femur	Ilium Sacrum Coccyx
Gluteus medius	Abducts *thigh*	Femur	Ilium
Gluteus minimus	Abducts *thigh*	Femur	Ilium
Adductors	Adduct *thigh*	Femur	Pubic bone
Hamstring group	Flexes *lower leg* Helps extend *thigh*	Tibia Fibula	Ischium Femur
Quadriceps femoris group, including rectus femoris	Extends *lower leg* Helps flex *thigh*	Tibia	Ilium Femur

outline summary

MUSCLE TISSUE

1 Structure and types
- a Striated muscles; also called skeletal muscle or voluntary muscle
- b Branching muscle; also called cardiac muscle
- c Nonstriated muscle; also called smooth muscle, visceral muscle, or involuntary muscle

2 Function—muscle cells specialize in contraction (shortening)

SKELETAL MUSCLES (ORGANS)

1 Structure
- a Composed mainly of striated muscle cells (fibers) and connective tissue
- b Most muscles extend from one bone across movable joint to another bone
- c Parts of a skeletal muscle
  - Origin—attachment to relatively immovable bone
  - Insertion—attachment to bone that moves
  - Body—main part of muscle
- d Muscles attach to bones by tendons—strong cords of fibrous connective tissue; some tendons enclosed in synovial-lined tubes, lubricated by synovial fluid; tubes called tendon sheaths; inflammation of tendon sheaths—tenosynovitis

e Bursae—small synovial-lined sacs containing small amount synovial fluid; located between some tendons and underlying muscles

2 Functions

a Muscles produce movement—as muscle contracts it pulls insertion bone nearer origin bone; movement occurs at joint between origin and insertion

b Groups of muscles usually contract to produce a single movement

Prime mover—muscle whose contraction is mainly responsible for producing a given movement

Synergists—muscles whose contractions help the prime mover produce a given movement

c Normal muscle functioning depends upon normal functioning of various other structures, notably nerves and joints

d Not all muscle contractions produce movements; isometric contractions increase the tension in muscles without producing movements; isotonic contractions produce movements; tonic contractions, or muscle tone, produce no movement but increase firmness (tension) of muscles that maintain posture, that is, the positions of parts of body; good posture necessary for best functioning—lessens fatigue, helps prevent bone and joint deformities

e Mobility (exercise) absolutely essential for healthy survival—necessary for maintaining muscle tone for normal functioning of heart and lungs, normal structure and functioning of bones and joints, and so on

3 Types of movements produced by skeletal muscle contractions

a Flexion—making angle at joint smaller

b Extension—making angle at joint larger

c Abduction—moving a part away from midline

d Adduction—moving a part toward midline

SPECIFIC FACTS ABOUT CERTAIN KEY MUSCLES

See Tables 3-1 and 3-2

POSTURE

1 Posture means position of body parts

2 Good posture important for many reasons—for example, to prevent fatigue and bone and joint deformities

3 Skeletal muscles maintain posture by counteracting the pull of gravity—by maintaining tone (partial contraction)

MUSCLE DISORDERS

1 Muscular dystrophy—progressive wasting and weakening of muscles

2 Paralysis—loss of ability to produce voluntary movements

3 Atrophy—decrease in muscle size

4 Hypertrophy—increase in muscle size

new words

abduction	isotonic
adduction	muscular dystrophy
atrophy	muscle tone
body of muscle	origin (of muscle)
bursae	paralysis
bursitis	posture
contraction	striated
extension	synovial fluid
flexion	synovial membrane
hypertrophy	tendon
insertion (of muscle)	tendon sheath
isometric	tenosynovitis

review questions

1 Compare the three kinds of muscle tissue as to location, microscopic appearance, and nerve control.

2 Explain why skeletal muscle functions are so important. What are the general functions of skeletal muscle?

3 Explain the terms flexion, extension, abduction, and adduction. Give an example of each.

4 Explain how skeletal muscles, bones, and joints work together to produce movements.

5 Why can a spinal cord injury be followed by muscle paralysis?

6 Can a muscle contract very long if its blood supply is shut off? Give a reason for your answer.

7 How do isometric contractions differ from isotonic contractions?

8 The correct term to substitute in the expression "bending your knee" is (extending? flexing?) your lower leg.

9 What is the name of the main muscle that

a Flexes the upper arm?

b Flexes the lower arm?

c Flexes the thigh?

d Flexes the lower leg?

e Extends the upper arm?

f Extends the lower arm?

g Extends the thigh?

h Extends the lower leg?

10 Give the approximate location of each of the following muscles and tell what movement it produces:

biceps brachii	pectoralis major
hamstrings	quadriceps femoris group
deltoid	latissimus dorsi

11 Describe brief changes that gradually take place in bones, joints, and muscles if a person habitually gets too little exercise.

12 Explain the following terms:

posture	hypertrophy
muscle tone	muscle tone
atrophy	good posture

unit three

Systems that control body functions

4

The nervous system

Any group of individuals who work together on anything have to have a boss. If their many separate jobs are going to accomplish one complex task, then obviously what each worker does must be controlled or regulated. This principle holds true regardless of the nature of the complex task to be done. Suppose the task is that of giving good care to hospitalized sick people. Can you imagine what would happen if there were no way of letting each individual worker know just what he was to do as his part of the job of caring for the hospital's patients? Pretty clearly, chaos would result. Some system for communicating with individual workers and controlling them is absolutely essential in any large organization.

The normal body accomplishes a gigantic and enormously complex job—that of keeping itself alive and healthy. Each one of its billions of individual cells performs some function that is a part of this big function. Control of the body's billions of cells is accomplished mainly by two communication systems, namely, the nervous system and the endocrine system. Both systems transmit information from one part of the body to another, but they do it in different ways. The nervous system transmits information by means of nerve impulses conducted from one nerve cell to another. The endocrine system transmits information by means of chemicals that are secreted by ductless glands into the bloodstream and are circulated from the glands to other parts of the body. Communication makes control possible. Nerve impulses and hormones communicate information to body structures—increasing or decreasing their activities, as needed for healthy survival. In other words, the communication systems of the body are also its control and integrating systems. Integrating systems are devices that weld the body's hundreds of different functions into its one overall function of keeping itself alive and healthy. We shall discuss the nervous system in this chapter and the endocrine system in Chapter 5. Our plan for this chapter is to relate basic information about the special cells of the nervous system and follow that with a discussion of the nervous system's organs.

Cells of the nervous system

Special cells found only in the nervous system are called *neurons,* or *nerve cells,* and *neuroglia. Neurons* specialize in the function of transmitting impulses. *Neuroglia* are special types of connective tissue cells. One reason for mentioning neuroglia is that one of the commonest types of brain tumor—called the *glioma*—develops from it. Neuroglia vary in size and shape. Some are relatively large cells that look somewhat like stars because of the many threadlike extensions that jut out from their surfaces. (These neuroglia are called *astrocytes,* a word that means star-shaped cells.) Their threadlike branches attach to both neurons and small blood vessels, holding these structures close to each other. Another kind of neuroglia, called *microglia* because of their small size, move about in brain tissue if it becomes inflamed. They act as microbe-eating scavengers. Microglia surround the microbes, draw them into their cytoplasm, and digest them. *Phagocytosis* is the scientific name for this important cellular process. You can read more about it in Chapter 8.

Each neuron consists of three main parts: a main part called the *neuron cell body,* one or more branching projections called *dendrites,* and one elongated projection known as an *axon.* Dendrites are the processes (projections) that transmit impulses to the neuron cell bodies, and axons are the processes that transmit impulses away from them.

Classified according to the direction in which they transmit impulses, there are three types of neurons: sensory neurons, motoneurons, and interneurons. *Sensory neurons* transmit impulses to the spinal cord and brain from all parts of the body. *Motoneurons* transmit impulses in the opposite direction—away from the brain and cord, not to them. Motoneurons do not conduct impulses to all parts of the body but only to two kinds of tissue—muscle and glandular epithelial. *Interneurons* conduct impulses from sensory neurons to motoneurons. Sensory neurons are also called afferent neurons; motoneurons are called efferent neurons; interneurons are called internuncial, intercalated, central, or connecting neurons.

Reflex arcs

Every moment of our lives, nerve impulses speed over neurons to and from our spinal cords and brains. If all impulse conduction ceases, life itself ceases. Only neurons can provide the rapid communication between the body's billions of cells, a process necessary for maintaining life. Chemical "messages" are the only other kind of communication the body can send and these travel much more slowly than impulses. They can move from one part of the body to another only by way of the circulating blood. And compared with impulse conduction, circulation is a very slow process indeed.

Nerve impulses can travel over literally millions of routes—routes made up of neurons, because neurons are the cells that conduct impulses. Hence, the routes traveled by nerve impulses are sometimes

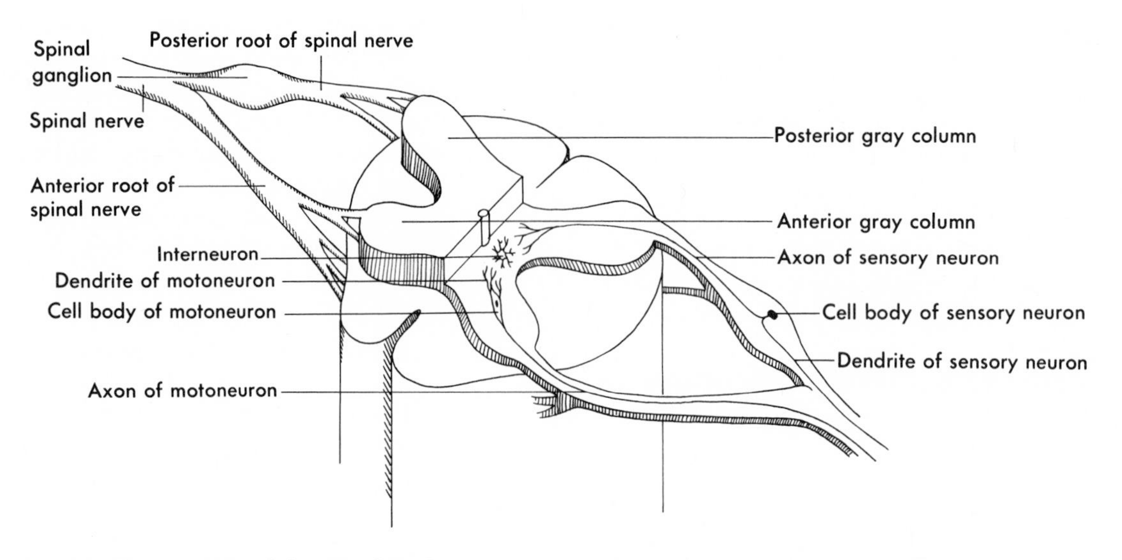

Fig. 4-1 Cross section of the spinal cord. Its interior is composed of gray matter surrounded by white matter. The left side of the diagram shows macroscopic structures only. The right side of the diagram shows locations of the dendrite, cell body, and axon (all microscopic structures) of a sensory neuron, an interneuron, and a motoneuron, the structures that compose a three-neuron reflex arc.

spoken of as neuron pathways. Their scientific name, however, is *reflex arcs*. The simplest kind of reflex arc is a *two-neuron arc*, so called because it consists of only two types of neurons—sensory neurons and motoneurons. *Three-neuron arcs* are the next simplest kind. They, of course, consist of all three kinds of neurons—sensory neurons, interneurons, and motoneurons. Reflex arcs are like one-way streets. They allow impulse conduction in only one direction. The next paragraph describes this direction in detail. Look frequently at Figure 4-1 as you read it.

Impulse conduction normally starts in receptors. *Receptors* are the beginnings of dendrites of sensory neurons. None are shown in Figure 4-1 because they are located at some distance from the cord—in skin and mucous membranes, for example. From receptors, impulses travel the full length of the sensory neuron's dendrite, through its cell body and axon, to the branching brushlike ends of the axon. The ends of the sensory neuron's axon contact interneurons. Actually, a microscopic space separates the axon endings of one neuron from the dendrites of another neuron. This space is called a *synapse*. After crossing the synapse, the impulses continue along the dendrites, cell bodies, and axons of the interneurons. Then they cross another synapse. Finally, they travel over the dendrites, cell bodies, and axons of motoneurons to a structure called an *effector* (because it "puts into effect" the message brought to it by nerve impulses). Effectors are either muscles or glands. Figure 4-1 does not show the effector in which the axon of the motoneuron ends, but it is a skeletal muscle. When impulses reach skeletal muscle, they cause its cells to contract. And this brings us to another definition. The response to impulse conduction over reflex arcs—in this case muscle contraction—is called a *reflex*. In short, impulse conduction by a reflex arc causes a reflex to occur. Muscle contractions and gland secretion are the only two kinds of reflexes.

Figure 4-1 reveals a number of other important facts. Dendrites, cell bodies, and axons of neurons are microscopic structures that are located in macroscopic structures. By comparing the right and left sides of the diagram, you can see, for example, that the dendrite of the sensory neuron lies in a spinal nerve. All sensory dendrites are relatively long structures because they extend from the skin and other organs all the way to the spinal cord. The cell bodies of sensory neurons lie outside of the cord or brain in structures called ganglia. The sensory cell body shown in Figure 4-1 is located in a spinal ganglion just outside the cord. *Spinal ganglia* are small swellings on the posterior roots of spinal nerves. Each one contains hundreds of sensory neuron cell bodies. Interneurons of spinal cord reflex arcs lie entirely within the central gray matter of the cord. In other words, their dendrites, cell bodies, and axons all lie in the inner core of the cord.

Dendrites and cell bodies of all motoneurons whose axons terminate in skeletal muscle are located in the anterior horns (columns) of the gray matter of the spinal cord—a fact of great practical importance. For instance, it explains how poliomyelitis causes paralysis. The polio virus attacks the anterior gray horns, injuring and destroying the motoneuron cell bodies located in them. They cannot then conduct impulses. And since they are the only motoneurons that send axons to skeletal muscles, the muscles supplied by the destroyed anterior horn motoneurons receive no nerve impulses; therefore, they cannot contract. They become paralyzed, in other words.

The term "reflex center" means the center of a reflex arc. As we have seen, the first part of a reflex arc consists of sensory neurons, and the last part consists of motoneurons. In spinal cord arcs, the sensory neurons conduct impulses to the cord, and the motoneurons conduct them away from the cord and out to muscles. The reflex centers, then, of all spinal cord arcs lie in spinal cord gray matter. In three-neuron arcs, the reflex centers consist of interneurons. In two-neuron arcs they are simply the synapses between sensory and motoneurons. We might define a reflex center as the place in a reflex arc where incoming impulses become outgoing impulses.

Organs of the nervous system

The organs of the nervous system are the spinal cord, the brain, and the numerous nerves of the body. Because the brain and spinal cord occupy a midline, or central, location in the body, together they are called the *central nervous system*, or simply the CNS. Similarly, the usual designation for the nerves of the body is PNS, meaning the *peripheral nervous system*—an appropriate name because nerves extend to outlying or peripheral parts of the body.

Spinal cord

The spinal cord occupies a protected location inside the spinal column. Nervous tissue is not a sturdy tissue. Even moderate pressure can kill nerve cells, so nature safeguards the chief organs made of this tissue—the spinal cord and the brain—by surrounding them with a tough, fluid-containing membrane (the meninges), and then by surrounding the membrane with bones. The *spinal meninges* form a tubelike covering around the spinal cord and line the vertebrae. Look at Figure 4-2 and

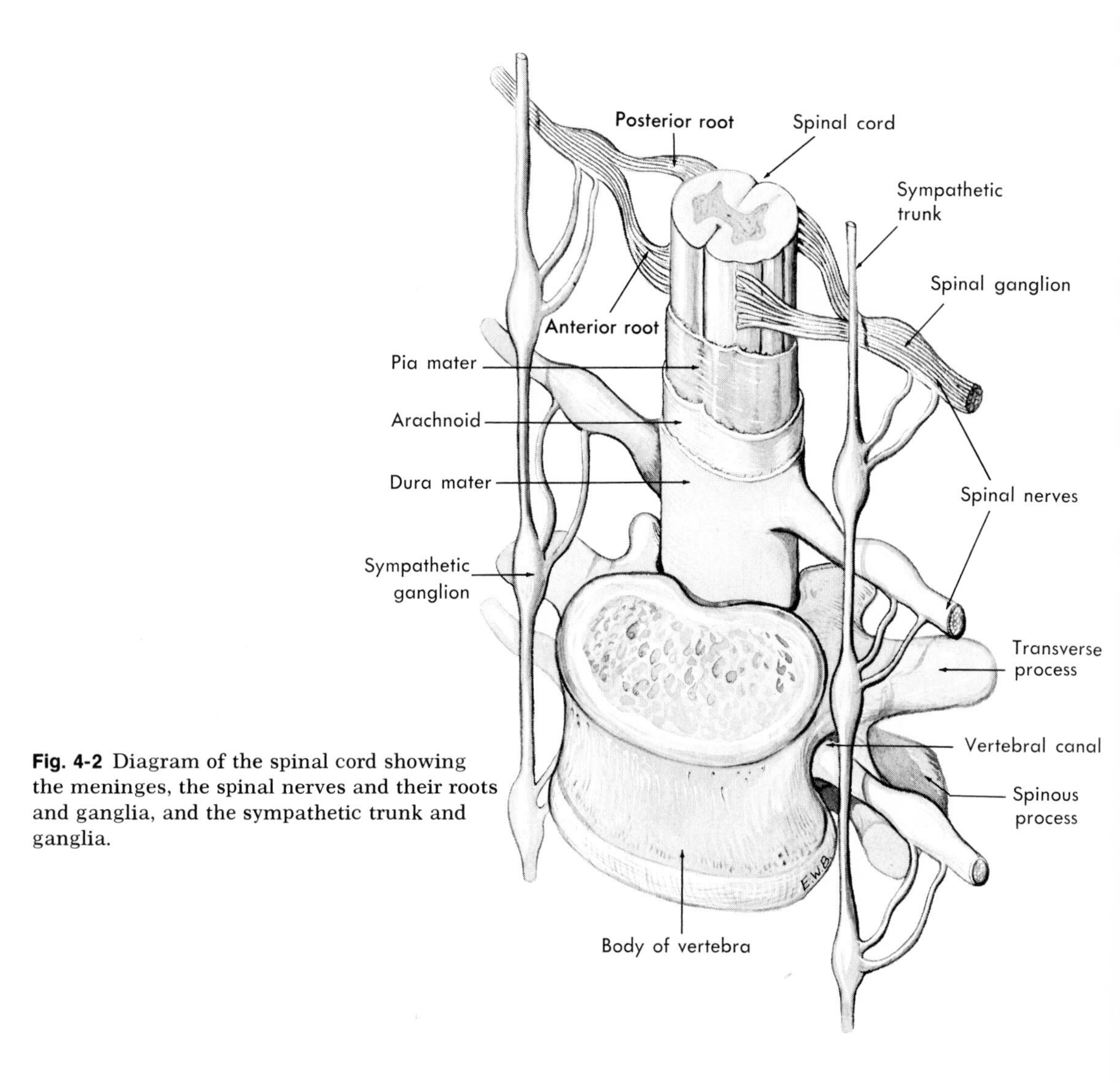

Fig. 4-2 Diagram of the spinal cord showing the meninges, the spinal nerves and their roots and ganglia, and the sympathetic trunk and ganglia.

you can identify the three layers of the spinal meninges. They are the dura mater (lines the vertebrae), the pia mater (covers the cord), and the arachnoid membrane between them. This middle layer of the meninges, the arachnoid membrane, resembles a cobweb with fluid filling in its spaces. (The word arachnoid means cobweblike. It comes from Arachne, the name of the girl who was changed into a spider because she boasted of the fineness of her weaving. At least, so an ancient Greek myth tells us.)

If you are of average height, your spinal cord is about 17 or 18 inches long. It lies inside the spinal column in the spinal cavity and reaches from the occipital bone down to the bottom of the first lumbar vertebra. (Place your hands on your hips and they will line up with your fourth lumbar vertebra; your spinal cord ends shortly above this level.) The spinal meninges,

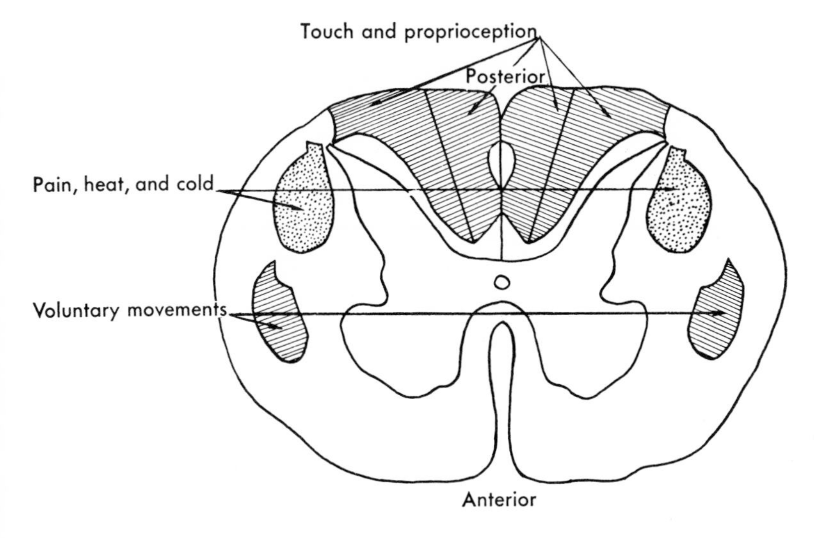

Fig. 4-3 Cross-sectional diagram of the spinal cord, showing the location of some important tracts (bundles of nerve fibers) that conduct impulses for the functions listed. The posterior white columns conduct impulses for touch and proprioception (sense of the position or movement of body parts); the lateral spinothalamic tracts, those for pain, heat, and cold; and the lateral corticospinal tracts, those for voluntary movements.

however, continue down almost to the end of the spinal column—an important fact for a physician to know, because it means that he can do a lumbar puncture without fear of damaging the cord.

Did you notice the H-shaped core of the spinal cord in Figures 4-1 and 4-2? It consists of gray matter and is composed mainly of dendrites and cell bodies of neurons. Bundles of axons, known as *tracts,* form columns of white matter around the gray matter.

To try to understand spinal cord functions, let us start by thinking about a hotel telephone switchboard. Suppose a guest in Room 108 calls the switchboard operator, asks for Room 520, and, in a second or so, someone in that room answers. Very briefly, three events took place: a message traveled into the switchboard, a connection was made in the switchboard, and a message traveled out from the switchboard. The telephone switchboard provided the connection that made possible the completion of this call. Or we might say that it transferred the incoming call to an outgoing line. Our spinal cords function similarly. They contain the centers for thousands and thousands of reflex arcs. Look back at Figure 4-1. The interneuron shown there is an example of a spinal cord reflex center. It switches, or transfers, incoming sensory impulses to outgoing, or motor, impulses, thereby making it possible for a reflex to occur. Reflexes that result from conduction over arcs whose centers lie in the cord are called *spinal cord* reflexes. Two common kinds of spinal cord reflexes are the *withdrawal reflexes* and the *jerk reflexes.* Conduction by three-neuron arcs produces a withdrawal reflex—for instance, pulling one's hand away from a hot surface. Conduction by two-neuron arcs produces a jerk reflex—for instance, the familiar knee jerk.

In addition to functioning as the primary reflex center of the body, the spinal cord also functions as a two-way conduction path for the brain. Figure 4-3 shows important nerve fiber bundles—its tracts, that is—located in the cord. Some conduct sensory impulses up to the brain; others conduct motor impulses down from the brain. Therefore, if an injury cuts the cord all the way across, impulses can no longer travel to the brain from any parts of the

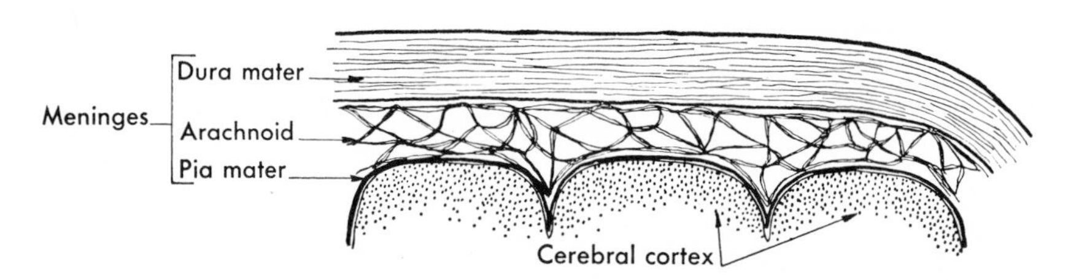

Fig. 4-4 Schematic drawing showing the structure of the meninges around the brain.

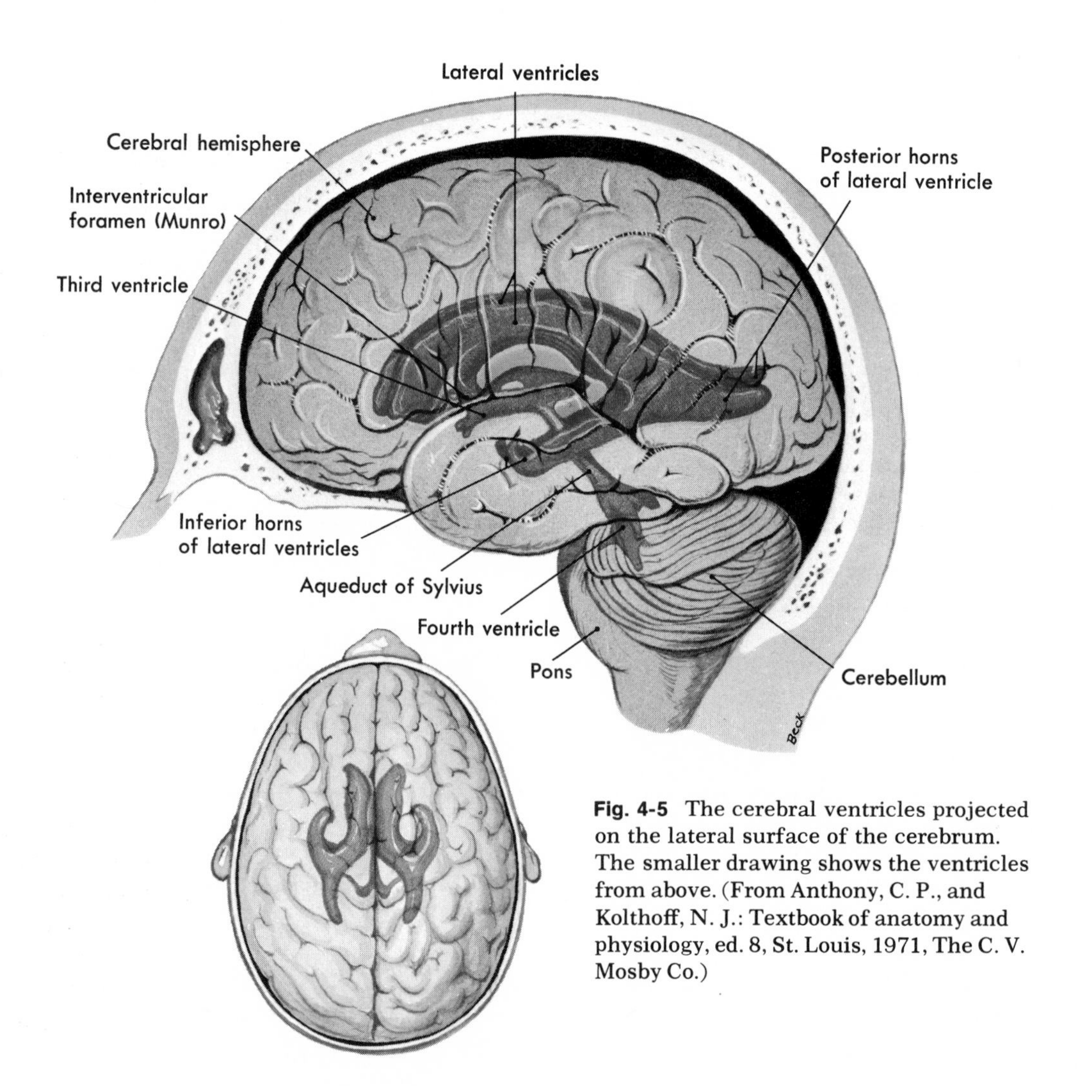

Fig. 4-5 The cerebral ventricles projected on the lateral surface of the cerebrum. The smaller drawing shows the ventricles from above. (From Anthony, C. P., and Kolthoff, N. J.: Textbook of anatomy and physiology, ed. 8, St. Louis, 1971, The C. V. Mosby Co.)

body located below the injury, and impulses cannot come from the brain down to such parts. In short, spinal cord injury can produce both a loss of sensation (anesthesia) and a loss of the ability to make voluntary movements (paralysis).

Brain

The brain consists of several parts, the most prominent of which are the cerebrum, the cerebellum, the midbrain, the pons, and the medulla. (Together, the midbrain, pons, and medulla seem to form a stem for the rest of the brain. In fact, they are often designated jointly as the *brainstem.*)

Two less prominently located parts of the brain, the thalamus and the hypothalamus, have captured a great deal of attention in recent years. They are the largest of several structures which together make up the *diencephalon,* the area of the brain tucked in between the cerebrum and the brainstem.

The meninges that form the protective covering around the spinal cord also extend up and around the brain to completely enclose it (Figure 4-4). Fluid fills the arachnoid spaces of the brain meninges as well as those of the spinal cord. This fluid, then, is *cerebrospinal fluid,* not just spinal fluid.

In Figure 4-5 you can see what irregular shapes the ventricles of the brain have. This illustration can also help you visualize the location of the ventricles if you remember two things—that these large spaces lie deep inside the brain and that there are two lateral ventricles. One lies inside the right half of the cerebrum (the largest part of the human brain), and one lies inside the left half of the cerebrum.

Cerebrospinal fluid is one of the body's circulating fluids. It forms continually from fluid moving out of the blood in brain capillaries and into the ventricles. From the lateral ventricles cerebrospinal fluid seeps into the third ventricle and circulates through the aqueduct of Sylvius (find this in Figure 4-5) into the fourth ventricle. From the fourth ventricle it seeps into the central canal of the cord and into the subarachnoid spaces. Then it moves leisurely down and around the cord and brain (in the subarachnoid spaces of their meninges) and returns to the blood (in veins of the brain). The exact circulation route of cerebrospinal fluid may well be information you will never need to know. But remembering that this fluid forms continually from blood, circulates, and is reabsorbed into blood can be useful. It can help you understand certain abnormalities you may see. Suppose a person has a brain tumor that presses on the aqueduct of Sylvius. This blocks the way for the return of cerebrospinal fluid to the blood. Since the fluid continues forming but cannot drain away, it accumulates in the ventricles or in the meninges (subarachnoid spaces around the brain). Other conditions besides brain tumors can cause an accumulation of cerebrospinal fluid in the ventricles. An example is *hydrocephalus,* or "water-on-the-brain."

MEDULLA. The medulla, the most vital part of the entire brain, is the only part of the brainstem we shall discuss. It is a bulb-shaped extension of the spinal cord and lies just inside the cranial cavity, right above a large hole (the foramen magnum) in the occipital bone. Like the spinal cord, the medulla consists of both gray and white matter. But their arrangement differs in the two organs. In the medulla, bits of the gray matter mix closely and intricately with the white matter to form what is called the *reticular formation* (reticular means netlike).

The medulla is the most vital part of the brain because it contains the vital centers. These consist of several clusters of neurons that serve as the centers of the reflex arcs that control three vital functions: respirations, the heartbeat, and blood pressure. Their individual names are the respiratory center, cardiac center, and vasomotor center. If their neurons fail to function (perhaps because they receive too little oxygen, or because they receive too much pressure from a brain tumor or skull injury), death follows quickly.

THALAMUS. The term "thalamus" is singular, suggesting the presence of one thalamus, but actually we have two of these organs, a right thalamus and a left thalamus. The right thalamus consists of a rounded mass of gray matter located deep inside the right half of the cerebrum; the left thalamus is a similar mass inside the left cerebral hemisphere. The thalamus consists chiefly of dendrites and cell bodies of neurons whose axons extend to various sensory areas of the cerebral cortex.

The thalamus functions to help produce sensations. Its neurons constitute one part of the neuron pathways that transmit the impulses resulting in sensations. All sensory impulses (except perhaps those responsible for the sense of smell) are conducted to the thalamus by neurons whose axons terminate in the thalamus. Here they synapse with neurons whose dendrites and cell bodies lie in the thalamus and whose axons extend to a sensory area in the cerebral cortex. Sensory impulses are conducted to the thalamus from below, and its neurons relay them to the cerebral cortex. In other words, the thalamus functions as a relay station for sensory impulses.

The thalamus almost surely is the part of the brain that associates sensations with emotions. Each of us experiences sensations with some degree of pleasantness or unpleasantness. Some sensations we find extremely pleasant, some unpleasant, and many others—perhaps the majority—we have very little feeling about one way or the other. For some reason, these feelings seem to come from the arrival of sensory impulses in the thalamus.

HYPOTHALAMUS. Translated literally, the word hypothalamus means "under the thalamus." Portions of the hypothalamus are located under each thalamus, and other parts lie anterior to the midbrain. The stalk (stem) and the posterior portions of the pituitary gland and the mammillary bodies are parts of the hypothalamus, which are located in front of the midbrain (Figures 4-6 and 4-7).

The old adage, "don't judge by appearance," applies well to an appraisal of the importance of the hypothalamus. Measured by size, the hypothalamus is one of the least significant parts of the brain, but measured by its contribution to healthy survival, the hypothalamus is an extremely important structure.

Some clusters of neurons in the hypothalamus act as centers for controlling the autonomic nervous system. Thus, the hypothalamus helps control most of our internal organs. It controls such things as the heart's beating, the constriction or dilation of the blood vessels, the contractions of the intestine, and the emptying of the bladder. The hypothalamus controls hormone secretion by both the anterior and posterior pituitary glands. Indirectly, therefore, it also helps control the amount of hormones secreted by practically all of the other endocrine glands. The hypothalamus acts as the center for controlling appetite, and for this reason plays an important part in regulating how much food an individual eats and how much he weighs. The hypothalamus functions in some way to keep us awake. Evidence of this is the fact that a person with disease or injury of the hypothalamus sleeps continually and is difficult to awaken.

Finally, the hypothalamus contains reward and punishment centers (also called

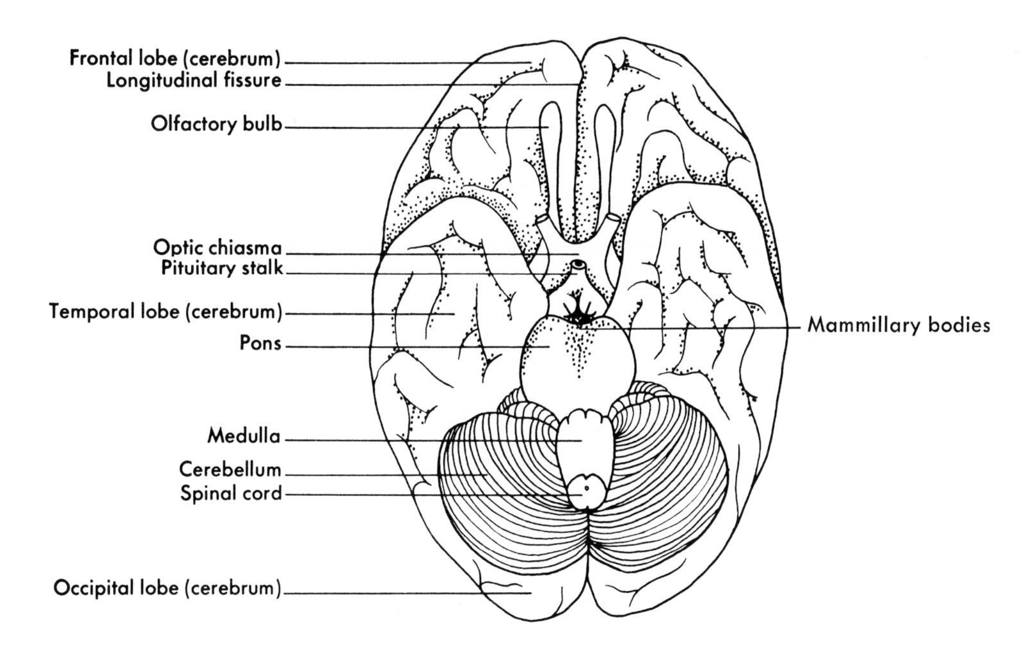

Fig. 4-6 Undersurface of the human brain. Note the location of two parts of the hypothalamus – the stalk of the pituitary gland and the mammillary bodies. (The pituitary gland has been cut off from its stalk.)

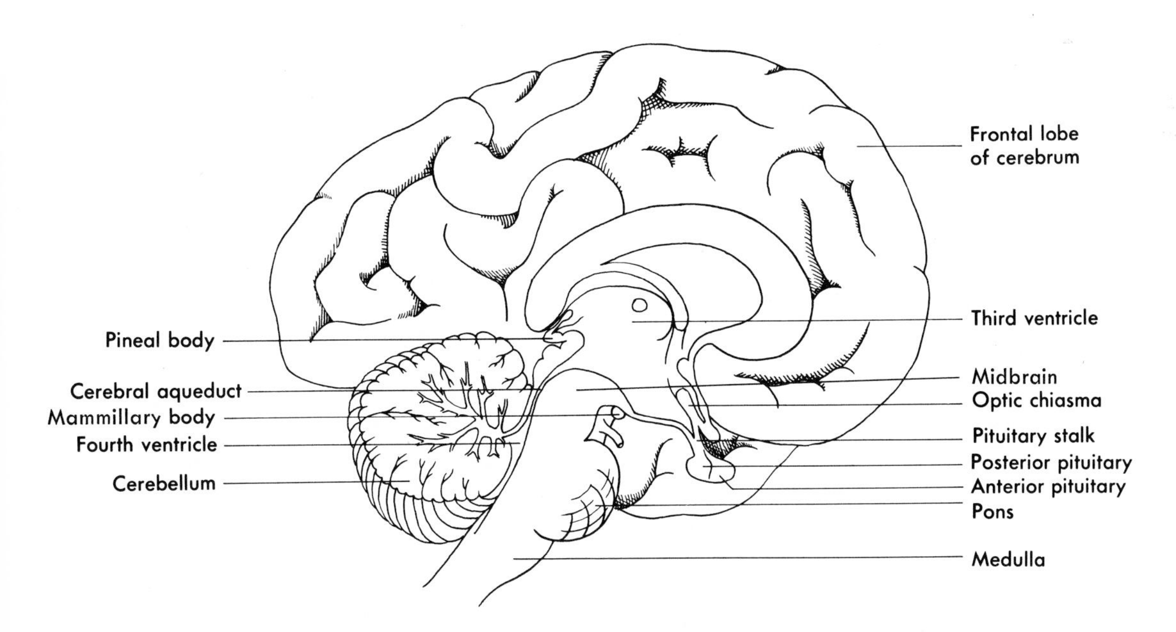

Fig. 4-7 Longitudinal section of the brain. You can find three parts of the hypothalamus in this diagram: the pituitary stalk, the posterior pituitary, and the mammillary bodies.

pleasure and displeasure centers), and so functions to produce some of our basic urges or drives. Experiments performed on rats and other animals in recent years have revealed some exciting information about this. For example, rats pressed levers hundreds of times an hour when these levers were arranged so as to stimulate tiny electrodes implanted in certain parts of the thalamus. That these animals must have found this stimulation pleasant seems an understatement. In fact, even when they were hungry and had not had food for many hours, they would still go on pressing the lever rather than stop to eat food when it was set before them. Other experiments seem to indicate that the hypothalamus also contains punishment or displeasure centers and that both reward and punishment centers exist in several other parts of the brain as well.

CEREBELLUM. Look at Figures 4-5, 4-6, and 4-7 to find the location, appearance, and size of the cerebellum. The cerebellum is the second largest part of the human brain. It lies under the occipital lobe of the cerebrum. Another fact about the cerebellum, but one not visible in the figures you have just examined, is that gray matter composes its outer layer and white matter composes the bulk of its interior.

Most of our knowledge about cerebellar functions has come from observing patients who have some sort of disease of the cerebellum and animals who have had the cerebellum removed. From such observations we know that the cerebellum plays an essential part in the production of normal movements. Perhaps a few examples will make this clear. A patient who has a tumor of the cerebellum frequently loses his balance and topples over; he may reel like a drunken man when he walks. He probably cannot coordinate his muscles normally. He may complain, for instance, that he is clumsy about everything he does – that he cannot even drive a nail or draw a straight line. With the loss of normal cerebellar functioning, he has lost the ability to make precise movements. The general functions of the cerebellum, then, are to produce smooth coordinated movements, maintain equilibrium, and sustain normal postures.

CEREBRUM. The largest part of the brain, as you can see in Figures 4-5, 4-6, and 4-7, is the cerebrum. A deep groove, the longitudinal fissure, divides it into two hemispheres, connected only in their lower middle portion. The outer layer of the cerebral hemispheres is called the cerebral cortex. (The word cortex means bark, or outer layer.) Many rounded ridges (convolutions) with grooves between them form the cerebral cortex. Like the cerebellar cortex, the cerebral cortex consists of gray matter. (In a living person the gray matter has a distinct pinkish cast.) White matter makes up most of the interior of the cerebrum.

Different areas of the cerebral cortex are called *lobes*. Frontal, parietal, temporal, and occipital are the names of the main lobes. As you might guess, the two frontal lobes lie under the frontal bones, the parietal lobes lie under the parietal bones, and so on.

What functions does the cerebrum perform? This is a hard question to answer briefly, because the neurons of the cerebrum do not function alone. They function with many other neurons located in many other parts of the brain and in the spinal cord. Neurons of these structures are continually bringing impulses to cerebral neurons and continually transmitting impulses away from them. If all other neurons were functioning normally and only cerebral neurons were not functioning, here are some of the things that you could not do. You could not think or use your will. You could not decide to make the smallest

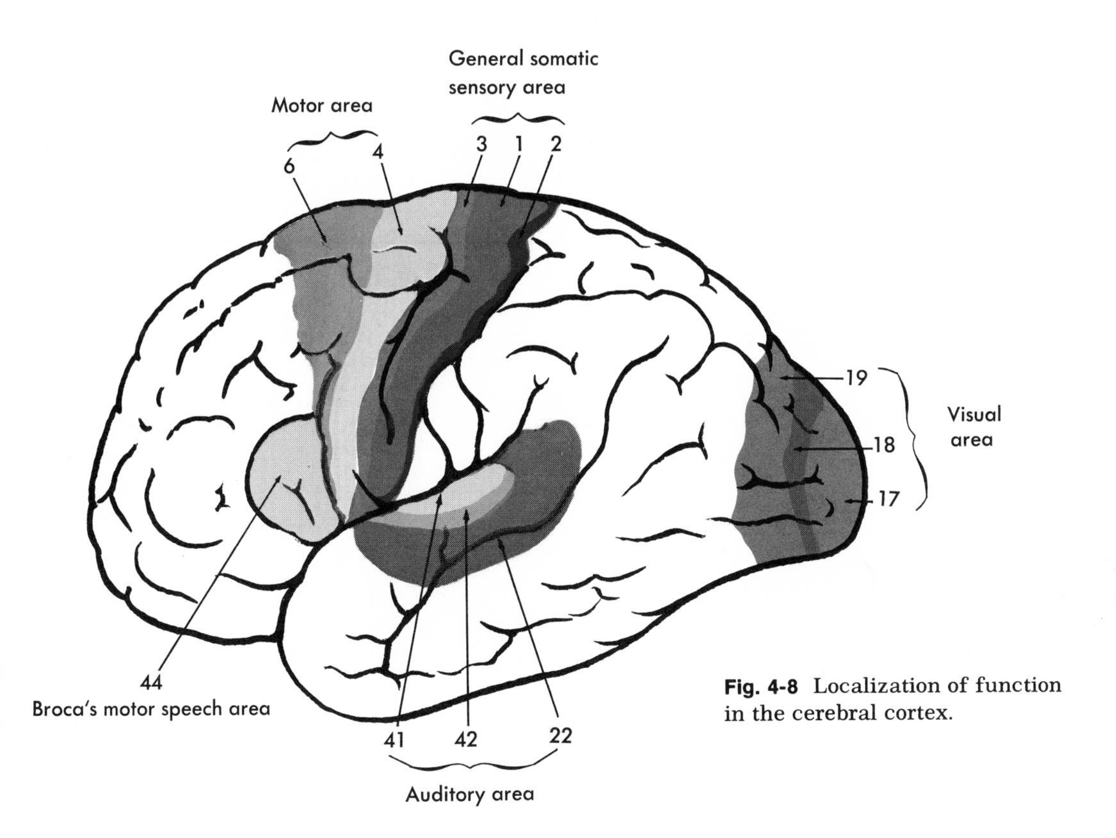

Fig. 4-8 Localization of function in the cerebral cortex.

movement or make it. You could not see or hear. You would not experience any of the other sensations that make life so rich and varied. Nothing would anger you or frighten you, and nothing would bring you great joy or great sorrow. In fact, you would be unconscious. The following five terms, then, summarize cerebral functions: consciousness, mental processes, sensations, emotions, and voluntary movements. Figure 4-8 shows the main functions of a few areas of the cerebral cortex.

Injury or disease can destroy neurons. All too often a hemorrhage from cerebral blood vessels (stroke) makes neurons of the motor area unable to function. When this happens, the victim can no longer make any voluntary movements. The motor area lies in the frontal lobe of the cerebrum. Based on your examination of Figure 4-8, in what lobe is the visual area located? The auditory area?

Spinal nerves

Thirty-one pairs of nerves attach to the spinal cord in the following order: eight pairs attach to the cervical segments, twelve pairs attach to the thoracic segments, five pairs attach to the lumbar segments, five pairs attach to the sacrospinal segments, and one pair attaches to the coccygeal segment. Unlike cranial nerves, spinal nerves have no special names; instead, a letter and number identifies each one. "C-1," for example, indicates the spinal nerves attached to the first segment of the cervical part of the cord; "T-8" indicates those attached to the eighth segment of the thoracic part of the cord, and so on.

Spinal nerves conduct impulses between the spinal cord and parts of the body not supplied by cranial nerves. As you can see in Figure 4-1, all spinal nerves contain both sensory and motor fibers. Spinal nerves, therefore, function to make possible both sensations and movements. This means that a patient with a disease or injury that damages a spinal nerve will not feel anything in the part of the body that nerve supplies, nor will he be able to

Table 4-1. Cranial nerves

*Nerve**	*Conducts impulses*	*Functions*
I. Olfactory	From nose to brain	Sense of smell
II. Optic	From eye to brain	Vision
III. Oculomotor	From brain to eye muscles	Eye movements
IV. Trochlear	From brain to external eye muscles	Eye movements
V. Trigeminal (or trifacial)	From skin and mucous membrane of head and from teeth to brain; also from brain to chewing muscles	Sensations of face, scalp, and teeth; chewing movements
VI. Abducens	From brain to external eye muscles	Turning eyes outward
VII. Facial	From taste buds of tongue to brain; from brain to face muscles	Sense of taste; contraction of muscles of facial expression
VIII. Auditory (or acoustic)	From ear to brain	Hearing; sense of balance
IX. Glossopharyngeal	From throat and taste buds of tongue to brain; also from brain to throat muscles and salivary glands	Sensations of throat, taste, swallowing movements; secretion of saliva
X. Vagus	From throat, larynx, and organs in thoracic and abdominal cavities to brain; also from brain to muscles of throat and to organs in thoracic and abdominal cavities	Sensations of throat, larynx, and of thoracic and abdominal organs; swallowing, voice production, slowing of heartbeat, acceleration of peristalsis
XI. Spinal accessory	From brain to certain shoulder and neck muscles	Shoulder movements; turning movements of head
XII. Hypoglossal	From brain to muscles of tongue	Tongue movements

*The first letters of the words of the following sentence are the first letters of the names of cranial nerves: "On Old Olympus' Tiny Tops A Finn and German Viewed Some Hops." Many generations of students have used this or a similar sentence to help them remember the names of cranial nerves.

move the part. Disease or injury of another part of the nervous system may also cause loss of sensation and voluntary movement. Do you recall what structure this is? If you do not, reread p. 59.)

Cranial nerves

Twelve pairs of cranial nerves attach to the undersurface of the brain. Their fibers conduct impulses between the brain and various structures in the head and neck and in the thoracic and abdominal cavities. For instance, the second cranial nerve (the optic nerve) conducts impulses from the eye to the brain, where these impulses produce the sensation of vision. The third cranial nerve (oculomotor nerve) conducts impulses from the brain to certain muscles of the eye, where these impulses cause contractions that move the eye. The names of each of the cranial nerves and a brief description of their functions are listed in Table 4-1.

Autonomic (involuntary) nervous system

The term "autonomic nervous system" is somewhat misleading. It suggests a separate and independent system of the body, but this is not so. The autonomic nervous system is a part of the body's one big complex nervous system. Autonomic neurons make up the autonomic nervous system. An autonomic neuron is one that conducts impulses out from either the brainstem or the spinal cord to either smooth muscle, cardiac muscle, or glandular epithelial tissue. These three kinds of tissues are called *visceral effectors,* and nerve impulses reach them only by way of autonomic neurons. In contrast, nerve impulses travel to skeletal muscles, or somatic effectors, by way of *somatic motoneurons.* Somatic effectors are the ones we can control voluntarily. Visceral effectors are the ones that we cannot control voluntarily. Except for a few rare individuals, we cannot, for example, will the smooth muscle in our intestines to contract more rapidly to increase peristalsis, or our heartbeats to speed up or to slow down, or our glands to increase or decrease their secretions. One way to define the autonomic nervous system, then, is as the part of the nervous system that controls the involuntary or automatic functions of the body, the processes most responsible for maintaining life.

The autonomic nervous system has two divisions, known as the *sympathetic nervous system* and the *parasympathetic nervous system.* When impulses over sympathetic fibers assume control of automatic functions, rapid changes take place within our bodies. Table 4-2 indicates some of these changes. The heart beats faster; most blood vessels constrict, causing blood pressure to shoot up; blood vessels in the skeletal muscles dilate to furnish them with more blood; sweat glands and adrenals secrete more abundantly; salivary and other digestive glands secrete more sparingly, and peristalsis becomes sluggish. These changes get us ready for strenuous muscular work or, as someone has so vividly stated, sympathetic impulses prepare us for "flight or fight." Therefore, the sympathetic nervous system can be thought of as the emergency nervous system. It takes over control of many of our internal functions when we are angry, frightened, worried, or when we are exercising hard—in short, whenever we are undergoing any kind of stress. If you observe patients before surgery or observe yourself, for that matter, when you are terribly frightened or have been running fast, you will probably notice some of the typical effects of sympathetic functioning just mentioned. Under ordinary circumstances, when we are

Table 4-2. Autonomic functions

Structure	*Parasympathetic control*	*Sympathetic control*
Heart muscle	Slows heartbeat	Accelerates heartbeat
Smooth muscle		
of most blood vessels	None	Constricts blood vessels
of blood vessels in skeletal muscles	None	Dilates blood vessels
of digestive tract	Increases peristalsis	Decreases peristalsis; inhibits defecation
of anal sphincter	Inhibits → opens sphincter for defecation	Stimulates → closes sphincter
of urinary bladder	Stimulates → contracts bladder	Inhibits → relaxes bladder
of urinary sphincters	Inhibits → opens sphincter for urination	Stimulates → closes sphincter
of eye		
iris	Stimulates circular fibers → constriction of pupil	Stimulates radial fibers → dilatation of pupil
ciliary	Stimulates → accommodation for near vision (bulging of lens)	Inhibits → accommodation for far vision (flattening of lens)
of hairs (pilomotor muscles)	No parasympathetic fibers	Stimulates → "goose pimples"
Glands		
adrenal medulla	None	Increases epinephrine secretion
sweat glands	None	Increases sweat secretion
digestive glands	Increases secretion of digestive juices	Decreases secretion of digestive juices

not struggling with mental or physical trials, parasympathetic control of most of our automatic functions dominates.

Study Table 4-2 to learn how certain body functions differ according to whether sympathetic or parasympathetic impulses dominate their control.

Figure 4-9 reveals an important difference between impulse conduction from the cord to somatic effectors (skeletal muscles) and from the cord to visceral effectors (smooth muscle, heart, glands). Examine the left side of the figure first. As you can see, one somatic motoneuron conducts impulses all the way from the cord to a somatic effector. This does not hold true for visceral effectors. You can see on the right side of the drawing that a preganglionic neuron conducts impulses from the cord to a ganglion, not to an effector. Impulses must then cross a synapse in the ganglion and be conducted by a postganglionic neuron to a visceral effector.

Figure 4-10 contains two words that you

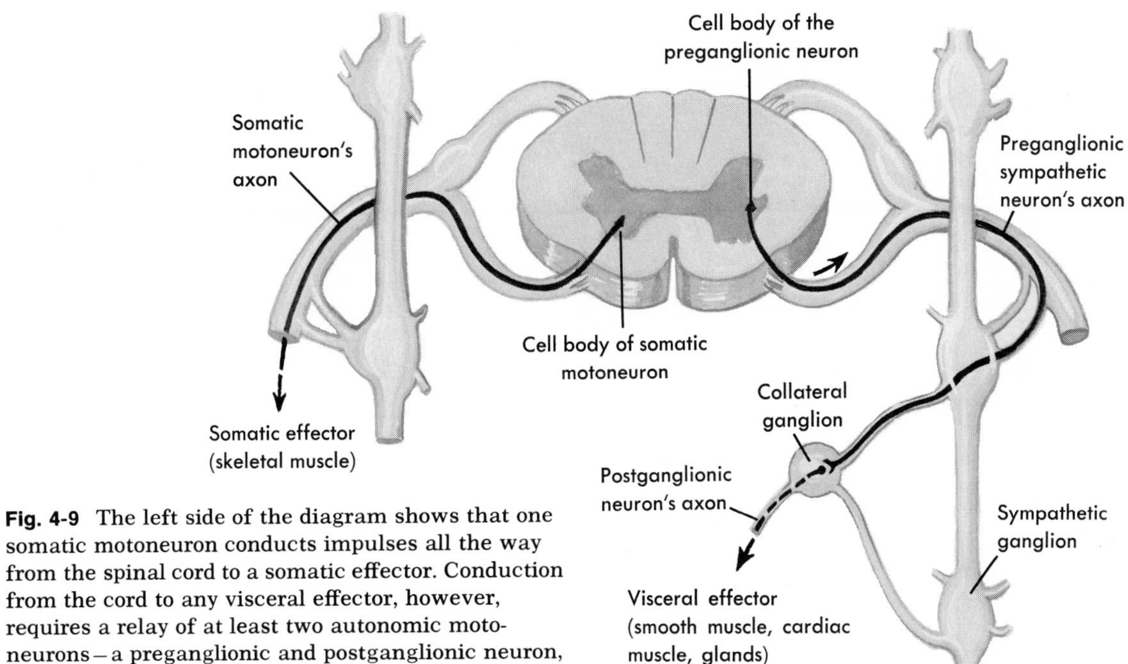

Fig. 4-9 The left side of the diagram shows that one somatic motoneuron conducts impulses all the way from the spinal cord to a somatic effector. Conduction from the cord to any visceral effector, however, requires a relay of at least two autonomic motoneurons—a preganglionic and postganglionic neuron, shown on the right side of the diagram.

may never have seen: acetylcholine and norepinephrine. These are chemical compounds and they perform enormously important functions. They make it possible for nerve impulses to cross synapses and neuroeffector junctions. The term "neuroeffector junction" suggests its meaning—the meeting place between a neuron and an effector. The part of a neuron that contacts an effector, as you already know, is its axon. But the axon ends in many fine branches, like bristles of a tiny brush. If you could have the opportunity to look at these axon endings magnified by an electron microscope, you would see a great many little bubblelike sacs in them. In the endings of some axons, these tiny sacs contain a few molecules of acetylcholine, and because they do, these axons are called *cholinergic fibers*. Now look again at Figure 4-10. Based on what you find there, decide whether this statement is true: the axons of all preganglionic autonomic neurons are cholinergic fibers. So far as is known, at any rate, this is true. What other cholinergic fibers do you find in Figure 4-10? Acetylcholine functions, then, as a "chemical transmitter" at autonomic synapses, at parasympathetic neuroeffector junctions, and also at somatic or skeletal muscle neuroeffector junctions (not shown in Figure 4-10). Norepinephrine functions as the chemical transmitter at most sympathetic neuroeffector junctions. Did you notice that the only place in Figure 4-10 where norepinephrine is shown is at the axon endings of a sympathetic postganglionic neuron. Axons secrete norepinephrine (noradrenaline) and are called *adrenergic fibers*. (Adrenergic comes from adrenaline, another name for epinephrine, a hormone secreted by the adrenal gland and a substance very similar to norepinephrine.)

You may be wondering how a chemical

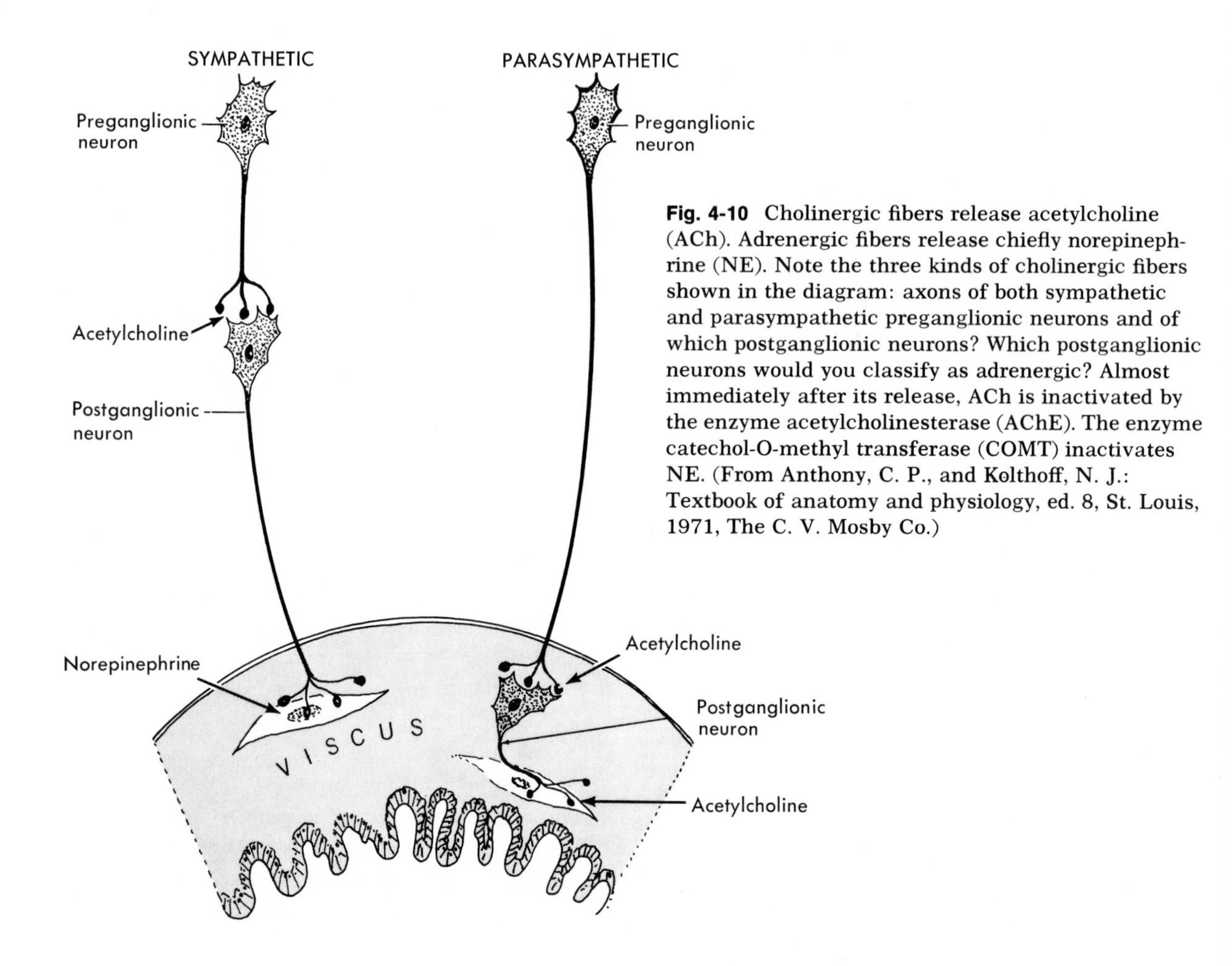

Fig. 4-10 Cholinergic fibers release acetylcholine (ACh). Adrenergic fibers release chiefly norepinephrine (NE). Note the three kinds of cholinergic fibers shown in the diagram: axons of both sympathetic and parasympathetic preganglionic neurons and of which postganglionic neurons? Which postganglionic neurons would you classify as adrenergic? Almost immediately after its release, ACh is inactivated by the enzyme acetylcholinesterase (AChE). The enzyme catechol-O-methyl transferase (COMT) inactivates NE. (From Anthony, C. P., and Kolthoff, N. J.: Textbook of anatomy and physiology, ed. 8, St. Louis, 1971, The C. V. Mosby Co.)

stored in sacs inside axon endings can transmit nerve impulses to another neuron or to an effector. Imagine first an impulse flashing down one of the preganglionic autonomic neurons shown in Figure 4-10. When the impulse reaches the axon's endings, it causes a number of their acetylcholine-containing sacs to release acetylcholine into the microscopic synaptic space. Here it contacts and stimulates the postganglionic neuron to start conducting impulses. Figure 4-11 summarizes this information in diagram form.

Sense organs

If you were asked to name the sense organs, what organs would you name? Can you think of any besides the eyes, ears, nose, and taste buds? Actually there are millions of other sense organs—all receptors are microscopic-size sense organs. Receptors, you will recall, are the beginnings of dendrites of sensory neurons.

Receptors are generously scattered about in almost every part of the body. To dem-

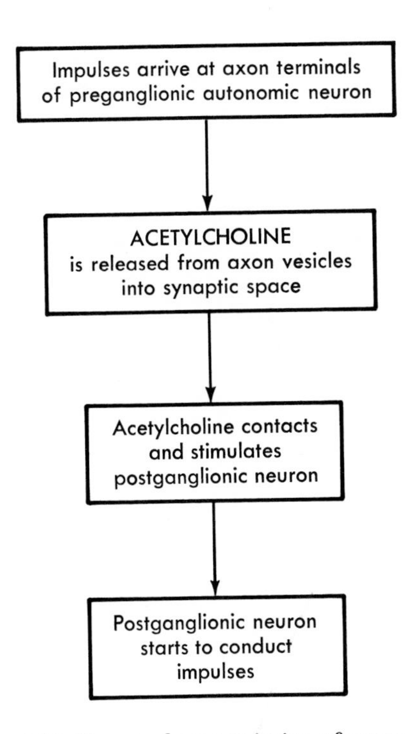

Fig. 4-11 Chemical transmission of nerve impulses across synapses in autonomic ganglia.

onstrate this fact, try pricking any point of your skin with a fine needle. You can hardly miss stimulating at least one receptor and almost instantaneously experiencing a sensation of mild pain. Stimulation of some receptors leads to the sensation of heat. Stimulation of other receptors gives the sensation of cold, and stimulation of still others gives the sensation of touch or pressure. When special receptors in the muscles and joints are stimulated, you sense the position of the different parts of the body and know whether they are moving or not and in which direction they are moving without even looking at them. Perhaps you have never realized that you have this sense of position and movement – a sense called *proprioception* or *kinesthesia*. Let us turn our attention now to two complex and remarkable sense organs – the eyes and ears.

Eye

When you look at a person's eye, you see only a small part of the whole eye. Three layers of tissue form the eyeball: the sclera, the choroid, and the retina. The outer layer, or *sclera*, consists of tough fibrous tissue. What we call the "white" of the eye is part of the front surface of the sclera. The other part of the front surface of the sclera is called the cornea and is sometimes spoken of as the "window" of the eye because of its transparency. At a casual glance, however, it does not look transparent but appears blue or brown or gray or green because it lies over the *iris*, the colored part of the eye. Mucous membrane known as the *conjunctiva* covers the entire front surface of the eyeball and lines both the upper and lower lids.

Two involuntary muscles make up the front part of the middle, or *choroid* coat, of the eyeball. One is the *iris*, the colored structure seen through the cornea, and the other is the *ciliary muscle* (Figure 4-13). What appears to be a black center in the iris is really a hole in this doughnut-shaped muscle; it is the *pupil* of the eye. Some of the fibers of the iris are arranged like spokes in a wheel. When they contract, the pupils dilate, letting in more light rays. Other fibers are circular. When they contract the pupils constrict, letting in fewer light rays. Normally, the pupils constrict in bright light and dilate in dim light.

The *lens* of the eye lies directly behind the pupil. It is held in place by a ligament attached to the ciliary muscle. When we look at distant objects, the ciliary muscle is relaxed, and the lens has only a slightly curved shape. To focus on near objects, however, the ciliary muscle must contract, causing the lens to become more bulging and more curved. Most of us become more

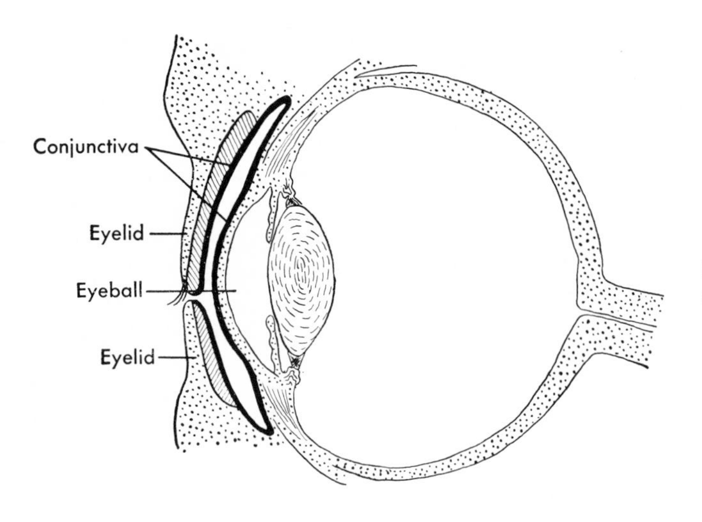

Fig. 4-12 Cross-sectional diagram of the eye showing location of the conjunctiva.

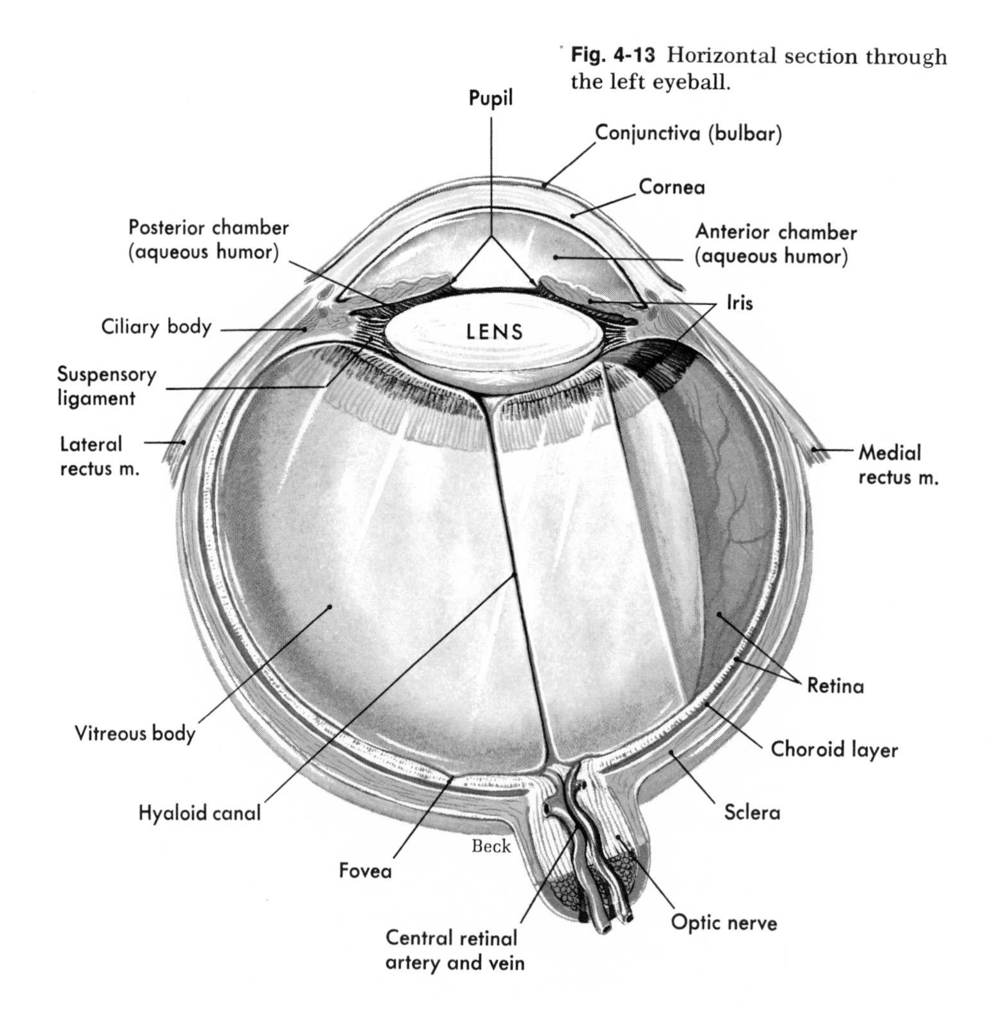

Fig. 4-13 Horizontal section through the left eyeball.

farsighted as we grow older. The reason we do is that our lenses lose their elasticity and can no longer bulge enough to bring near objects into focus. Presbyopia, or "oldsightedness," is the name for this condition.

The *retina,* or innermost coat of the eyeball, contains microscopic structures called rods and cones because of their shapes. Both are receptors for vision. Dim light can stimulate the rods, but fairly bright light is necessary to stimulate the cones. In other words, *rods* are the receptors for night vision and *cones* for daytime vision. (Cones are also the receptors for color vision.)

Fluids fill the hollow inside of the eyeball. They maintain the normal shape of the eyeball and help refract light rays; that is, the fluids bend light rays so as to bring them to a focus on the retina. *Aqueous humor* is the name of the fluid in front of the lens (in the anterior cavity of the eye) and *vitreous humor* is the name of the jellylike fluid behind the lens (in the posterior cavity).

Ear

The ear is much more than a mere appendage on the side of the head. A large part of the ear—and by far its most important part—lies hidden from view deep inside the temporal bone. Part of the external ear, all of the middle ear, and all of the internal ear are located here.

The *external ear* has two parts; the pinna (or auricle) and the ear canal (or external acoustic meatus). The *pinna* is the appendage on the side of the head. The *ear canal* is a curving tube in the temporal bone that leads from the pinna to the middle ear.

The *middle ear* (or tympanic cavity) is a tiny cavity hollowed out of the temporal bone. This cavity is lined with mucous membrane and contains three very small bones. The names of these ear bones are Latin words that describe their shapes—*malleus* (hammer), *incus* (anvil), and *stapes* (stirrup). The *tympanic membrane* (commonly called the eardrum) separates the middle ear from the external ear canal. The "handle" of the malleus attaches to the inside of the tympanic membrane, and the "head" attaches to the incus. The incus attaches to the stapes, and the stapes fits into a small opening, the *oval window,* that opens into the internal ear. A point worth mentioning, because it explains the frequent spread of infection from the throat to the ear, is the fact that a tube—the *auditory* or *eustachian tube*—connects the throat with the middle ear. The mucous lining of the middle ears, eustachian tubes, and the throat are extensions of one continuous membrane. Consequently, a sore throat may spread to produce a middle ear infection *(otitis media).* It can even cause a mastoid infection *(mastoiditis).* Mastoid spaces (sinuses) also open into the middle ear cavity. Because the mucous lining of the mastoid sinuses is also continuous with the mucous lining of the middle ear, it provides a direct route for a middle-ear infection to spread to produce *mastoiditis.*

The *internal ear* has two parts. One is made of bone, the other of a membrane that lies inside the bone. Both have complicated shapes, and for this reason are called *labyrinths.* Each labyrinth has three parts: *vestibule, semicircular canals,* and *cochlea* (Figure 4-14). The semicircular canals are three half circles; the cochlea is shaped like a snail shell, which is what the word cochlea means.

Two special sense organs for two different kinds of sensations—hearing and balance—are located in the internal ear. The hearing sense organ, which lies inside the cochlea, is called the *organ of Corti.* There are two balance, or equilibrium, sense organs—one called the *macula,*

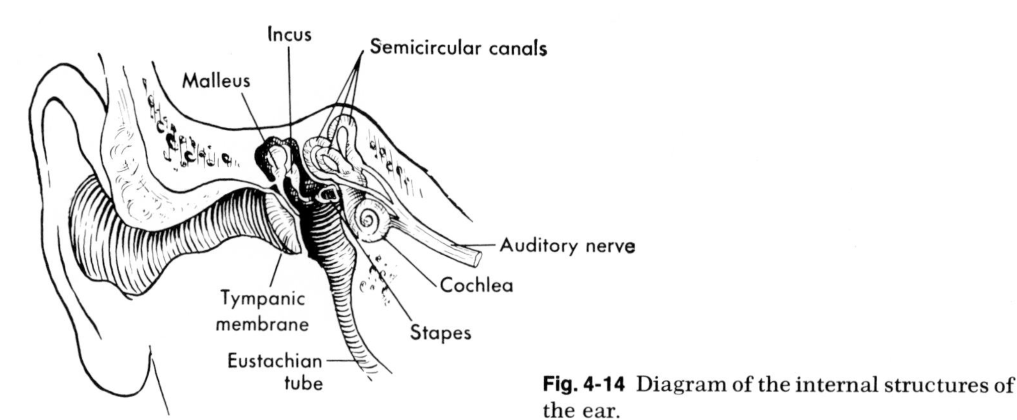

Fig. 4-14 Diagram of the internal structures of the ear.

which is located in the vestibule of the inner ear, and the other called the *crista*. The crista is located in the semicircular canals.

Diseases of the central nervous system

Although many diseases affect the brain and spinal cord, we shall mention only four: encephalitis, multiple sclerosis, Parkinson's disease, and strokes.

The first part of the word *encephalitis* means brain, and "itis" means inflammation. Inflammation of the brain presumably comes from a virus infection. Because victims of encephalitis sleep excessively, it is also called "sleeping sickness."

Sclerosis is the Greek word for hardness. In the disease *multiple sclerosis*, small hard patches develop in many scattered areas of the cord and brain. Symptoms vary according to the areas involved.

Parkinson's disease is a disease of the parts of the brain that function to produce our automatic movements and postures. Because shaking or tremors are common symptoms of this disease, it is also known as "shaking palsy." (Paralysis agitans is its scientific name.)

What lay people call *strokes*, or *apoplexy*, physicians call *cerebral vascular accidents*, or CVA's. The name cerebral vascular accident suggests an abnormality of blood vessels in the cerebrum. A clot may form in them, or they may rupture and hemorrhage. (*Cerebral hemorrhage* is another name for a stroke.) As a result of the clot or hemorrhage, neurons in the affected part of the brain become damaged or destroyed. The symptoms of a CVA vary. Sometimes they are few and unimportant; sometimes they are numerous and fatal. The outcome depends upon where in the cerebrum the hemorrhage occurs and how massive it is.

Diseases of nerves

Shingles, or herpes zoster, is a virus infection of a nerve. Little blisters develop on the skin along the course of the nerve.

Sciatica is an inflammation of the largest nerve of the body, the sciatic nerve in the thigh. Sciatica is one form of *neuritis* (nerve inflammation) and is a cause of *neuralgia* (nerve pain).

outline summary

CELLS OF NERVOUS SYSTEM

1 Neuroglia–connective tissue cells of two main types
  a Astrocytes–star-shaped cells that anchor small blood vessels to neurons
  b Microglia–small cells that move about in inflamed brain tissue carrying on phagocytosis, that is, they engulf and destroy microorganisms and other injurious particles
2 Neurons or nerve cells
  a Consist of three main parts: dendrites–conduct impulses to cell body of neuron; cell body of neuron; and axon–conducts impulses away from cell body of neuron
  b Neurons classified according to function as sensory–conduct impulses to spinal cord and brain; motoneurons–conduct impulses away from brain and cord out to muscles and glands; and interneurons–conduct impulses from sensory neurons to motoneurons

REFLEX ARCS

1 Nerve impulses are conducted from receptors to effectors over neuron pathways or reflex arcs; conduction by reflex arc results in a reflex, that is, contraction by a muscle or secretion by a gland
  b Simplest reflex arcs called two-neuron arcs–consist of sensory neurons synapsing in spinal cord with motoneurons; three-neuron arcs consist of sensory neurons synapsing in spinal cord with interneurons that synapse with motoneurons

ORGANS OF NERVOUS SYSTEM

1 Central nervous system (CNS)–brain and spinal cord
2 Peripheral nervous system (PNS)–all nerves

Spinal cord

1 Coverings–vertebrae and spinal meninges
2 Fluid spaces–subarachnoid spaces of meninges and central canal inside cord
3 Structure–outer part composed of white matter made up of many bundles of axons called tracts; interior composed of gray matter made up mainly of cell bodies of neurons
4 Functions
  a Serves as center for all spinal cord reflexes
  b Sensory tracts conduct impulses to brain and motor tracts conduct impulses from brain

Brain

1 Main divisions–medulla, pons, midbrain, hypothalamus, thalamus, cerebellum, and cerebrum
2 Coverings–cranial bones and meninges
3 Fluid spaces–subarachnoid spaces of brain meninges and ventricles inside brain
4 Medulla
  a Lowest part of brain; enlarged extension of spinal cord in cranial cavity
  b Most vital part of brain since vital centers located in medulla control heartbeat, diameter of blood vessels (and therefore blood pressure), and respirations
5 Thalamus
  a Rounded mass of gray matter in each half (hemisphere) of cerebrum; located on each side of third ventricle
  b Neurons in thalamus relay sensory impulses to cerebral cortex sensory areas
  c Functions some way to produce emotions of pleasantness or unpleasantness associated with sensations
6 Hypothalamus
  a Acts as the major center for controlling autonomic nervous system; therefore helps control the functioning of most internal organs
  b Controls hormone secretion by both anterior and posterior pituitary glands; therefore indirectly helps control hormone secretion by most other endocrine glands
  c Acts as center for controlling appetite; therefore helps regulate amount of food eaten and body weight
  d Functions in some way to maintain the waking state
  e Probably contains reward and punishment centers
7 Cerebellum
  a Second largest part of human brain
  b Helps control muscle contractions so that they produce coordinated movements so we can maintain balance, move smoothly, and sustain normal postures
8 Cerebrum
  a Largest part of human brain
  b Outer layer called cerebral cortex; made up of lobes, which are made up of convolutions; cortex composed mainly of cell bodies of neurons
  c Interior of cerebrum composed mainly of nerve fibers arranged in bundles called tracts
  d Functions of cerebrum–mental processes of all types, including sensations, consciousness, and voluntary control of movements

Nerves

1 Spinal nerves
  a Structure–contain dendrites of sensory neurons and axons of motor neurons
  b Functions–conduct impulses
    Necessary for sensations
    Necessary for voluntary movements
2 Cranial nerves
  See Table 4-1

AUTONOMIC (INVOLUNTARY) NERVOUS SYSTEM

1 Autonomic neurons are motoneurons that conduct impulses to visceral effectors (smooth muscle, cardiac muscle, and glands)
2 Divisions – sympathetic and parasympathetic; both sympathetic and parasympathetic neurons conduct impulses to heart and most smooth muscle and glands
3 Functions
  a In general, sympathetic and parasympathetic impulses tend to produce opposite effects (see Table 4-2); together, they help control heartbeat, contraction of smooth muscle, and secretion of glands
  b Sympathetic impulses dominate control of involuntary, automatic functions in times of stress, causing them to respond in ways that make possible maximum muscular activity; sympathetic impulses prepare body for "fight or flight"
  c Parasympathetic impulses dominate control of most involuntary functions under normal conditions

SENSE ORGANS

1 All receptors (beginning of dendrites of sensory neurons) are sense organs
2 Special sense organs – eyes and ears
3 Kinds of senses – many more than the familiar five senses; for example several kinds of touch senses; proprioception – sense of position and movement

Eye

1 Coats of eyeball
  a Sclera – tough outer coat; whites of eye; cornea is transparent part of sclera over iris
  b Choroid – front part of this coat made up of ciliary muscle and iris, the colored part of eye; pupil is hole in center of iris; contraction of iris muscle dilates or constricts pupil
  c Retina – innermost coat of eye; contains rods (receptors for night vision) and cones (receptors for day vision and color vision)
2 Conjunctiva – mucous membrane that covers front surface of eyeball and lines lids
3 Lens – transparent body behind pupil; focuses light rays on retina
4 Eye fluids
  a Aqueous humor – in anterior cavity in front of lens
  b Vitreous humor – in posterior cavity behind lens

Ear

1 Parts
  a External ear – consists of pinna and auditory canal
  b Middle ear – contains auditory bones (malleus, incus, and stapes); lined with mucous membrane; eustachian tubes and mastoid sinuses open into middle ear
  c Internal ear, or labyrinth – vestibule, semicircular canals, and cochlea are three divisions, organ of Corti, sense organ of hearing, lies in cochlea; sense organs of equilibrium are the macula and the crista; macula lies in vestibule and crista lies in semicircular canals

DISEASES OF THE NERVOUS SYSTEM

1 Encephalitis – inflammation of the brain
2 Multiple sclerosis – progressive disease in which small hard patches develop in many areas of the cord and brain, causing a variety of symptoms
3 Parkinson's disease (shaking palsy) – disease of certain parts of the brain (basal ganglia) that function to produce normal automatic movements and postures; common symptom is shaking, or tremors
4 Strokes (apoplexy, cerebral vascular accidents, or CVA) – blood clots in one or more cerebral blood vessels or rupture of vessels, cause injury of nearby brain areas and loss of their function, causing variety of symptoms
5 Shingles (herpes zoster) – virus infection of a nerve, manifested by appearance of blisters on the skin over the nerve
6 Sciatica – inflammation of the sciatic nerve, or neuritis; neuritis causes neuralgia, that is, nerve pain

new words

adrenergic fibers	neuroglia
anesthesia	neurons
axon	paralysis
cholinergic fibers	phagocytosis
dendrite	receptors
effectors	reflex arc
neuroeffector junctions	reflex

review questions

1 What general function does the nervous system perform?
2 What other system performs the same general function as the nervous system?
3 What general functions does the spinal cord perform?
4 What does "CNS" mean?
5 What are the meninges?
6 Why is the medulla considered the most vital part of the brain?
7 What general functions does the cerebellum perform?
8 What general functions does the cerebrum perform?
9 What general functions do spinal nerves perform?
10 What are some of the functions performed by cranial nerves?
11 Which pair or pairs of cranial nerves would you nickname "seeing nerves?" "hearing nerves?" "smelling nerves?" "tasting nerves?"
12 Would a person be blind, deaf, or neither, if both of his eighth cranial nerves atrophied?
13 What general functions does the autonomic nervous system perform?
14 Describe as fully as you can the structure of the eye.
15 Describe as fully as you can the structure of the ear.
16 What is each of the following?
conjunctiva iris
cornea retina
organ of Corti
17 Explain briefly why most old people are far-sighted.
18 Define briefly each of the terms listed under "new words."
19 Identify each of the following:
interneuron spinal ganglion
motoneuron reflex center
sensory neuron visceral motoneuron
somatic motoneuron

5

The endocrine system

Have you ever seen a giant or a dwarf? Have you ever known anyone who had sugar diabetes or a goiter? If so, you have had visible proof of the importance of the endocrine system for normal development and health.

The endocrine system performs the same general functions as the nervous system, namely, communication and control. The nervous system provides rapid, brief control via fast-traveling nerve impulses. The endocrine system provides slower but longer-lasting control via hormones (chemicals) secreted into and circulated by the blood.

The organs of the endocrine system are located in widely separated parts of the body—in the cranial cavity, in the neck, in the thoracic cavity, in the abdominal cavity, in the pelvic cavity, and outside the body cavities. Note the names and locations of the endocrine glands shown in Figure 5-1 and Table 5-1.

All the organs of the endocrine system are glands, but all glands are not organs of the endocrine system. Those glands that discharge their secretions into ducts are not endocrine glands. *Endocrine glands* are those that secrete chemicals known as *hormones* into the blood.

In this chapter you will read about the main functions of individual endocrine glands and will learn of the importance of the endocrine system for healthy survival.

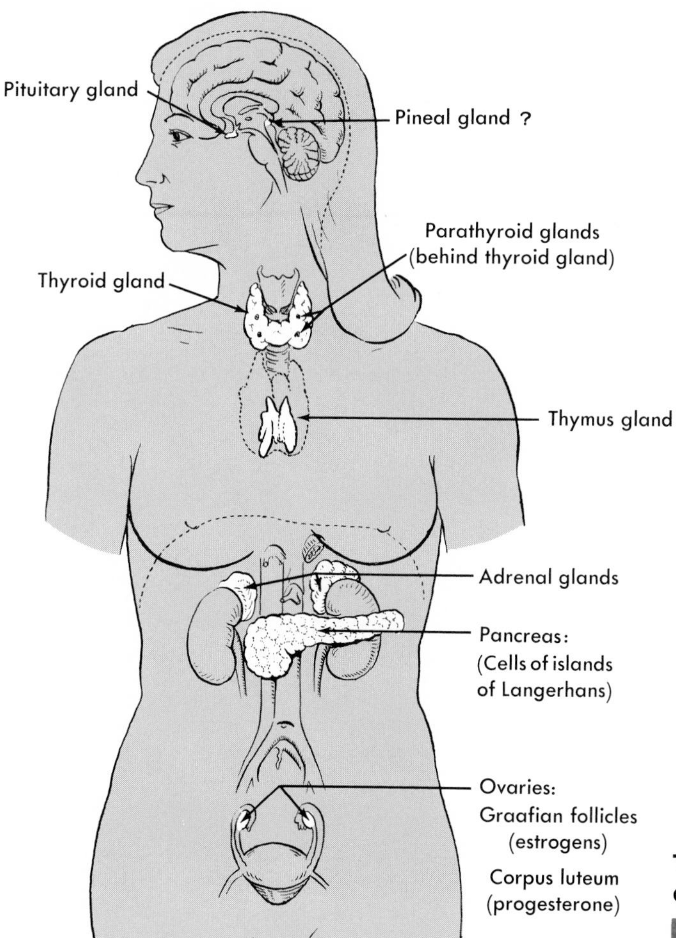

Fig. 5-1 Location of the endocrine glands in the female. Pineal body postulated but not proved to be an endocrine gland. Dotted line around thymus gland indicates maximum size at puberty. Which glands appear in this figure but not in Table 5-1? What gland does not appear in this figure but does appear in Table 5-1? (From Anthony, C. P., and Kolthoff, N. J.: Textbook of anatomy and physiology, ed. 8, St. Louis, 1971, The C. V. Mosby Co.)

Table 5-1. Names and locations of endocrine glands

Name	*Location*
Pituitary gland (hypophysis cerebri)	Cranial cavity
Anterior lobe (adenohypophysis)	
Posterior lobe (neurohypophysis)	
Thyroid gland	Neck
Parathyroid glands	Neck
Adrenal glands	Abdominal cavity (retroperitoneal)
Adrenal cortex	
Adrenal medulla	
Islands of Langerhans	Abdominal cavity (pancreas)
Ovaries	Pelvic cavity
Graafian follicle	
Corpus luteum	
Testes (interstitial cells)	Scrotum

Pituitary body (or hypophysis)

The pituitary body, an organ no larger than a pea, is really two endocrine glands. One is called the anterior pituitary gland (or adenohypophysis) and the other is called the posterior pituitary gland (or neurohypophysis). The protected location of the pituitary body suggests this organ's importance. The pituitary body lies buried deep in the cranial cavity, in the small depression of the sphenoid bone that is shaped like a saddle and is called Turk's saddle or the sella turcica. A stemlike structure, the pituitary stalk attaches the gland to the undersurface of the brain. More specifically, the stalk attaches the pituitary body to the hypothalamus.

Anterior pituitary gland hormones

The anterior pituitary gland secretes six major hormones. Five of them are usually referred to by their abbreviated names—TSH, ACTH, FSH, LH, and GH. Learning these is a little like learning a mixed-up alphabet, but if you associate the letters with the full names of the hormones (given in Figure 5-2), you probably will find the task fairly easy.

The anterior pituitary gland is often called the *master gland,* because four of its hormones act on four other endocrine glands to control both their structure and functioning. The four hormones that act in this way are the thyroid-stimulating hormone, adrenocorticotrophic hormone, follicle-stimulating hormone, and luteinizing hormone.

Thyroid-stimulating hormone (TSH) acts on the thyroid gland. It stimulates growth of the thyroid gland, and it also stimulates the secretion of thyroid hormone.

Adrenocorticotrophic hormone (ACTH) acts on the adrenal cortex. It stimulates growth of the adrenal cortex and secretion of cortisol (hydrocortisone) and similar substances.

Follicle-stimulating hormone (FSH) stimulates primary graafian follicles to start growing and to continue developing to maturity, that is, to the point of ovulation. FSH also stimulates follicle cells to secrete estrogens. In the male, FSH stimulates the seminiferous tubules to grow and form sperm.

Luteinizing hormone (LH) performs four functions. It acts with FSH to stimulate a follicle and ovum to complete their growth to maturity. It stimulates follicle cells to secrete estrogens. It causes ovulation (rupturing of the mature follicle with expulsion of its ripe ovum). Because of this function, LH is sometimes called the ovulating hormone. Finally, LH stimulates the formation of a golden body, the corpus luteum, in the ruptured follicle—the process is called luteinization. This function, obviously, is the one that earned LH its title of luteinizing hormone.

The other two important hormones secreted by the anterior pituitary gland are growth hormone and prolactin (lactogenic hormone). Growth hormone (GH) tends to speed up the movement of digested proteins (amino acids) out of the blood and into the cells, and this tends to accelerate the cells' anabolism of amino acids to form tissue proteins; hence, this action promotes normal growth. Growth hormone also affects fat and carbohydrate metabolism: it accelerates fat catabolism but slows glucose catabolism. This means that less glucose leaves the blood to enter cells and, therefore, the amount of glucose in the blood tends to increase. Thus, growth hormone and insulin have opposite effects on blood sugar. Insulin tends to decrease blood sugar, and growth hormone tends to increase it. Too much insulin in the blood produces *hypoglycemia* (lower than normal blood glucose concentration). Too

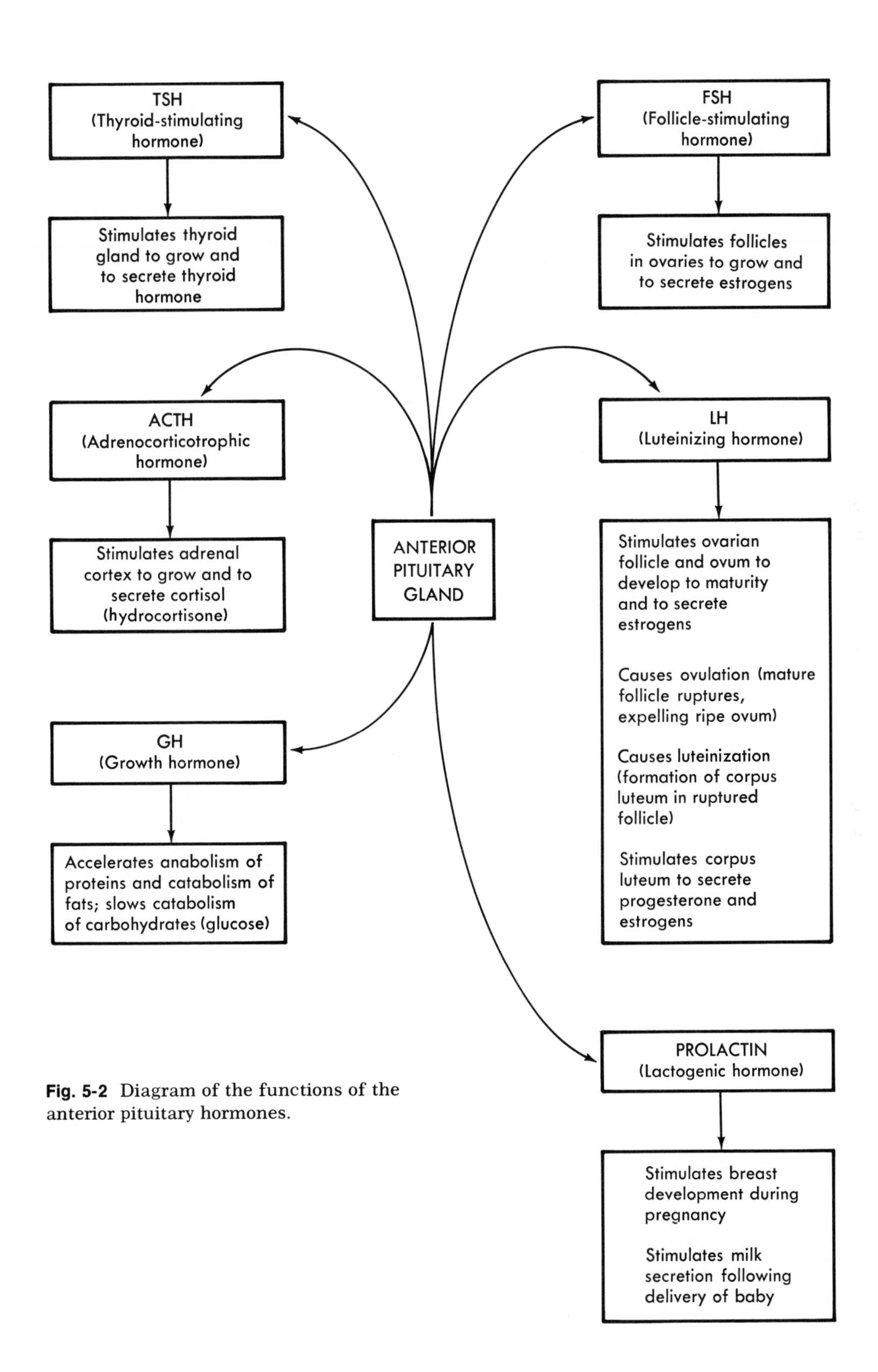

Fig. 5-2 Diagram of the functions of the anterior pituitary hormones.

much growth hormone produces *hyperglycemia* (higher than normal blood glucose concentration). This type of hyperglycemia is appropriately called pituitary diabetes, and growth hormone is referred to as a diabetogenic (diabetes-causing) or hyperglycemic hormone.

During pregnancy, prolactin stimulates the breast development necessary for eventual milk secretion, or lactation. Soon after the delivery of a baby, prolactin stimulates the breasts to start secreting milk, a function suggested by prolactin's other name, lactogenic hormone.

For a brief summary of anterior pituitary hormones and their functions, see Figure 5-2.

Posterior pituitary gland hormones

The posterior pituitary gland secretes two hormones—antidiuretic hormone (ADH) and oxytocin. ADH accelerates the reabsorption of water from urine in kidney tubules back into the blood. With more water moving out of the tubules into the blood, less water remains in the tubules, and therefore less urine leaves the body. This is the reason the name antidiuretic hormone is an appropriate one—"anti" means against and "diuretic" means increasing the excretion of urine. ADH acts against increasing urine. It decreases the amount of urine, in other words.

The posterior pituitary hormone oxytocin is secreted by a woman's body before and after she has a baby. Oxytocin stimulates contraction of the smooth muscle of the pregnant uterus, and so is thought to initiate and maintain labor. This is why physicians sometimes prescribe oxytocin injections to induce or to increase labor. Oxytocin also performs a function that is important to a newborn baby. It causes the glandular cells of the breast to release milk into ducts from which a baby can obtain it by sucking.

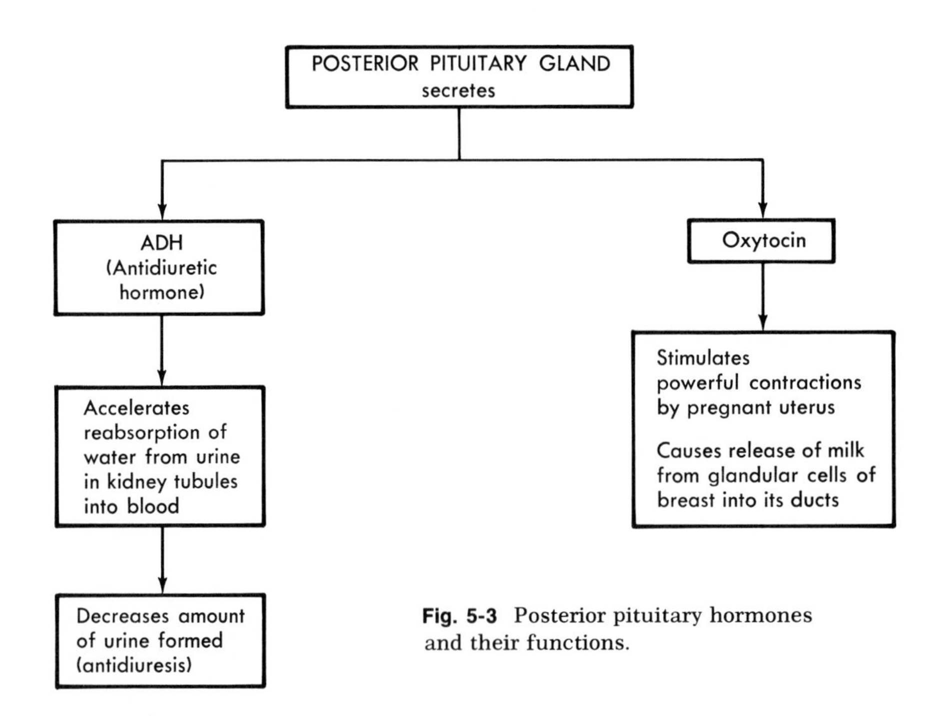

Fig. 5-3 Posterior pituitary hormones and their functions.

Thyroid gland

Earlier in this chapter we mentioned that some endocrine glands lie outside all of the body cavities. The thyroid is one of these. It lies in the neck just below the larynx.

The thyroid gland secretes two hormones — thyroid hormone and calcitonin (also called thyrocalcitonin).

The thyroid hormone influences every one of the billions of cells of our bodies. It makes them speed up their release of energy from foods. Thyroid hormone stimulates catabolism; in other words, it increases basal metabolism. This has far-reaching effects. Because all body functions depend upon a normal supply of energy, they all depend upon normal thyroid secretion. Even normal mental development and normal physical growth and development depend upon normal thyroid functioning.

If the thyroid secretes too little hormone, catabolism slows down. As a result, cells have too little energy and cannot do their work properly. If the thyroid secretes too much hormone, catabolism speeds up too much. This produces very noticeable effects. The hyperthyroid individual appears nervous, jumpy, and excessively active — hyperactive, we say.

Calcitonin tends to decrease the concentration of calcium in the blood. Exactly how it does this, no one has yet established. One idea about this process is that calcitonin may inhibit the breakdown of bone and thereby inhibit the release of calcium from bone into blood. Another idea is that it may increase the amount of calcium deposited in bone from the blood. Possibly, it may do both. It may decrease the amount of calcium moving into blood from bone and increase the amount of calcium moving out of blood into bone. Both actions would tend to decrease blood calcium concentration. In short, calcitonin functions to help maintain homeostasis of blood calcium. It prevents a harmful excess of calcium in the blood (hypercalcemia) from developing.

Parathyroid glands

The parathyroids are small glands. There are usually four of them, and they are found on the back of the thyroid. The parathyroid glands secrete *parathormone*.

Parathormone tends to increase the concentration of calcium in the blood — the opposite effect of the thyroid gland's calcitonin. Parathormone tends to increase bone breakdown, thereby increasing the release of calcium into the blood. This is a matter of life and death importance because our cells are extremely sensitive to changing amounts of blood calcium. They cannot function normally either with too much or with too little calcium. For example, with too much blood calcium, brain cells and heart cells soon do not function normally; a person becomes mentally disturbed and his heart may stop. However, with too little blood calcium, nerve cells become overactive — sometimes to such an extreme degree that they bombard muscles with so many impulses that the muscles go into spasms. This is called *tetany*.

Based on what you have just read about calcitonin and parathormone, which condition might lead to too little blood calcium and tetany — parathormone deficiency or parathormone excess?

Adrenal glands

As you can see in Figure 5-1, an adrenal gland curves over the top of each kidney. Although from the surface an adrenal gland appears to be only one organ, it actually is two separate endocrine glands, namely, the adrenal cortex and the adrenal medulla. Does this two-glands-in-one structure remind you of another endocrine organ? (See p. 78.) The adrenal cortex is the outer part of an adrenal gland and the medulla is its inner part. Adrenal cortex hormones have different names and quite different actions from adrenal medulla hormones.

Adrenal cortex

Hormones secreted by the adrenal cortex are called *corticoids*. There are three kinds of corticoids: glucocorticoids (or GC's, for short), mineralocorticoids (or MC's), and sex hormones.

Glucocorticoids (chiefly the hormone called cortisol or hydrocortisone) resemble thyroid hormone in that they act on all body cells. Glucocorticoids help control some of the most vital cellular functions. Their two most important general functions are these: glucocorticoids enable cells to carry on normal metabolism of all three kinds of foods (proteins, fats, carbohydrates); glucocorticoids also make cells, and therefore the body as a whole, able to resist stress. The following paragraphs explain these two general functions of glucocorticoids a little more fully.

Glucocorticoids in normal amounts act on cells in some way to make them able to carry on normal anabolism and catabolism of proteins, fats, and carbohydrates. Higher blood concentrations than usual of glucocorticoids accelerate both protein and fat catabolism. High blood concentrations of glucocorticoids especially accelerate the breakdown of tissue proteins to amino acids. These amino acids then move out of the cells into the blood and circulate to the liver. Liver cells change them to glucose by the process of *gluconeogenesis*. When this newly made glucose enters the blood, it tends to increase the blood sugar concentration above its normal level. Excess glucocorticoids, therefore, tend to produce hyperglycemia. Excess glucocorticoids also tend to cause tissue wasting, because they accelerate the breakdown (catabolism) of tissue proteins and fat. In fact glucocorticoids are often called "catabolic hormones." Do you recall the name of an anterior pituitary hormone that in excessive amounts tends to produce hyperglycemia? (See pp. 78 and 80.)

When extreme stimuli act on the body, they produce an internal state or condition known as *stress*. Surgery, hemorrhage, infections, severe burns, and intense emotions are examples of extreme stimuli that bring on stress. The normal adrenal cortex responds to the condition of stress by quickly increasing its secretion of glucocorticoids. Increased glucocorticoid secretion is only one of many responses the body makes to stress, but it is one of the first *stress responses* that takes place. It is also probably the most essential stress response, because directly or indirectly, a higher than normal blood concentration of glucocorticoids brings about a great many of the other stress responses. Among the most important of these are the following:

1. Increased catabolism of tissue proteins and fat, and after a period of time, tissue wasting and loss of body weight
2. Increased gluconeogenesis and high blood sugar (hyperglycemia)
3. Atrophy (decrease in size) of the thymus gland and the lymph nodes
4. A marked decrease in the number of lymphocytes and eosinophils in the blood
5. Decreased formation of antibodies and therefore less immunity to infectious diseases
6. Slower formation of scar tissue and therefore slower wound healing

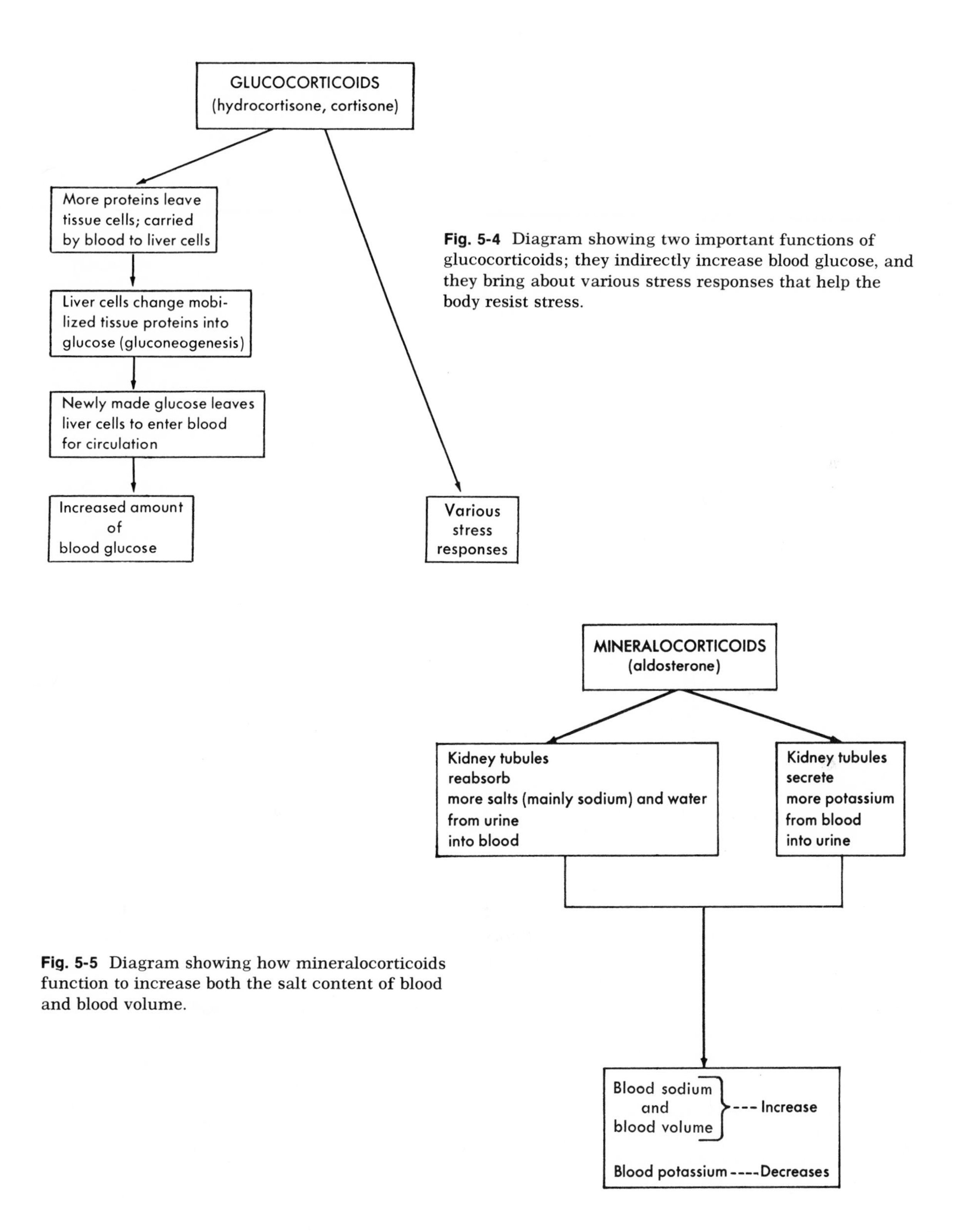

Fig. 5-4 Diagram showing two important functions of glucocorticoids; they indirectly increase blood glucose, and they bring about various stress responses that help the body resist stress.

Fig. 5-5 Diagram showing how mineralocorticoids function to increase both the salt content of blood and blood volume.

Mineralocorticoids perform different functions from glucocorticoids. As their name suggests, they help control the amount of certain mineral salts in the blood. Aldosterone is the name of the chief MC. Remember its main functions—to increase the amount of sodium and decrease the amount of potassium in the blood—for these changes lead to other profound changes. Aldosterone increases blood sodium and decreases blood potassium by influencing the kidney tubules to speed up their reabsorption of sodium back into the blood so that less of it will be lost in the urine. At the same time, the tubules increase their secretion of potassium so that more of this mineral will be lost in the urine. Aldosterone also tends to speed up kidney reabsorption of water.

One kind of sex hormone secreted by the adrenal cortex acts like the male sex hormone testosterone. Compounds of this type, therefore, are called *androgens* (from the Greek word "andros," meaning man). Androgens are male sex hormones; they produce masculinizing effects. Strangely enough, even a woman's adrenal glands secrete some androgens. Under certain circumstances, the adrenal cortex also secretes female sex hormones, but usually in such small amounts that they do not have a noticeable effect on the body.

Adrenal medulla

The adrenal medulla, or inner portion of the adrenal glands, secretes *epinephrine* and *norepinephrine.* (Biochemists refer to these compounds as *catecholamines.*)

Our bodies have many ways to defend themselves against enemies that threaten their well-being. A physiologist might state this same truth by saying that the body resists stress by means of many stress responses. We have just discussed increased glucocorticoid secretion. An even faster-acting stress response is increased secretion by the adrenal medulla. (It occurs very rapidly because nerve impulses conducted by sympathetic nerve fibers stimulate the adrenal medulla.) When stimulated, it literally squirts epinephrine and norepinephrine into the blood. Like the glucocorticoids, these hormones help the body resist stress, but unlike the glucocorticoids, they are not essential for maintaining life. Glucocorticoids, the hormones from the adrenal cortex, on the other hand, both help the body resist stress and are essential for life.

Suppose you suddenly faced some threatening situation. Imagine that you found a lump in your breast, or that your doctor told you that you had to have a dangerous operation, or that a gunman threatened to kill you. Almost instantaneously, the medullas of your two adrenal glands would be galvanized into feverish activity. They would quickly secrete large amounts of epinephrine (adrenalin) into your blood. Many of your body functions would seem to become supercharged. Your heart would beat faster; your blood pressure would rise; more blood would be pumped to your skeletal muscles; your blood would contain more sugar for more energy, and so on. In short, you would be geared for strenuous activity, for "fight or flight," to use the words of a famous physiologist, Walter Cannon. Epinephrine prolongs and intensifies changes in body function brought about by one division of the nervous system. (Do you recall which one? Check your answer on pp. 65 and 66.)

Islands of Langerhans

All of the endocrine glands discussed so far are big enough to be seen without a magnifying glass or microscope. The islands of Langerhans, in contrast, are too tiny to be seen without a microscope. These glands are merely little clumps of cells scattered like islands in a sea among the pancreatic cells that secrete the pancreatic digestive juice. Paul Langerhans discovered these cell islands in the pancreas almost one hundred years ago—hence their name islands of Langerhans.

Two kinds of cells in the islands of Langerhans are those called alpha and beta cells. Alpha cells secrete a hormone called *glucagon*, whereas beta cells secrete one of the most famous of all hormones, *insulin*.

Glucagon accelerates a process called *liver glycogenolysis*. Glycogenolysis is a chemical process by which the glucose stored in the liver cells in the form of glycogen is converted to glucose. This glucose then leaves the liver cells and enters the blood. Glucagon, therefore, tends to increase blood glucose concentration.

Insulin and glucagon serve as antagonists. In other words, insulin tends to decrease blood glucose concentration; glucagon tends to increase it. Insulin is the only hormone that can decrease blood glucose concentration. Several other hormones, however, tend to increase its concentration. We have already named three of these: glucocorticoids, growth hormone, and glucagon. Insulin decreases blood glucose by accelerating its movement out of the blood, through cell membranes, and into cells. As glucose enters the cells at a faster rate, the cells increase their metabolism of glucose. Insulin, then, functions to decrease blood glucose and to increase glucose metabolism.

If the islands of Langerhans secrete a normal amount of insulin, a normal amount of glucose enters the cells, and a normal amount of glucose stays behind in the blood. ("Normal blood sugar" is about 80 to 120 milligrams of glucose in every 100 milliliters of blood.) If the islands of Langerhans secrete too much insulin, as they sometimes do when a person has a tumor of the pancreas, then more glucose than usual leaves the blood to enter the cells and blood sugar decreases. If the islands of Langerhans secrete too little insulin, as they do in *diabetes mellitus*, less glucose leaves the blood to enter the cells so that blood sugar increases—sometimes to even three or more times the normal amount.

Thymus gland

One of the body's best-kept secrets has been the thymus gland's functions. Before 1961 about all that was known about this organ was that it was located in the mediastinal portion of the chest, that it grew in size until a child reached the age of puberty (see Figure 5-1), and that afterwards over the years it gradually became smaller. Since 1961, starting with some experiments done on newborn mice, many facts have been discovered and ideas suggested about the thymus gland's functions. It is now known that the thymus functions in some way to make a person able to develop immunity against various diseases.

Physiologists today think that the thymus probably does two things that are essential for the development of immunity. First, it probably produces the original lymphocytes formed in the body before birth and continues to produce them after birth. (Lymphocytes then travel from the thymus to the lymph nodes and spleen by way of the circulation.) Second, the thymus probably forms a hormone essential for immunity. Apparently this hormone must be present for a short time after a baby is born, if he is ever going to be able to become immune to any disease. Scientists have postulated that the thymus hormone acts on lymphocytes, causing them to change into plasma cells, which then form the antibodies that produce immunity.

Pineal body

Like the thymus gland, the pineal body has long been a mystery organ. Recently, some investigators have suggested that it may secrete a hormone (called melatonin) that serves as a kind of "biological clock" for regulating the ovaries' activity, but for now this is merely interesting speculation.

Female sex glands

A woman's sex glands are her two ovaries. Each ovary contains two different kinds of glandular structures, namely, the graafian follicles and the corpus luteum. *Graafian follicles* secrete estrogens, the "feminizing hormone." The *corpus luteum* secretes chiefly progesterone, but also some estrogens. We shall save our discussion of the structure of these endocrine glands and the functions of their hormones for Chapter 10.

Male sex glands

Some of the cells of the testes secrete semen (the male reproductive fluid) into ducts. Other cells, referred to as interstitial cells, secrete the male hormone testosterone into the blood. The interstitial cells of the testes, therefore, are the male endocrine glands. Testosterone is the "masculinizing hormone." Chapter 10 contains more information about the structure of the testes and the functions of testosterone.

Endocrine diseases

Diseases of the endocrine glands are numerous, varied, and sometimes spectacular. Tumors or other abnormalities frequently cause the glands to secrete either too much (hypersecretion) or too little (hyposecretion) of their hormones. We shall identify a few of the more common conditions characterized by hypersecretion or hyposecretion of hormones.

Excessive secretion of growth hormone during the early years of life produces a condition called *giantism*. The name suggests the obvious characteristics of this condition. The child grows to giant size. If the anterior pituitary gland secretes too much growth hormone after the growth years, then the disease called *acromegaly* develops. The individual's bones—particularly those of the face, hands, and feet—enlarge gradually but markedly, often changing the person's appearance enough to make him almost unrecognizable. Excess growth hormone tends to produce excess blood sugar (hyperglycemia or "pituitary diabetes").

Almost everyone has known someone with a hyperactive thyroid gland, or, as it is commonly called, a *goiter*. Among the chief symptoms of a hyperactive thyroid are an increased metabolic rate, a very rapid heartbeat, and nervous excitability.

Hyposecretion of thyroid hormone during the early years of childhood leads to a disease called *cretinism*. Malformed dwarfs are victims of cretinism. Their metabolic rate is low, and frequently they show signs of retarded mental, physical, and sexual development. If the thyroid functions normally during the growth years and then starts to secrete too little of its hormone, the disease *myxedema* results. Besides a low metabolic rate, a person with myxedema usually shows signs of lessened mental and physical vigor. They generally gain weight and often complain that their hair is falling out.

Cushing's syndrome is the medical name for the hypersecretion of glucocorticoids. For some reason many more women than men develop Cushing's syndrome. Its most noticeable features are the so-called moon face and buffalo hump that develop because of the redistribution of body fat.

A deficiency of insulin secretion leads to an excess of blood sugar, the identifying characteristic of *diabetes mellitus*, probably the best-known of all endocrine disorders. A higher than normal blood sugar in turn leads to many other symptoms, and if high enough for long can cause death.

Deficient secretion of the antidiuretic hormone by the posterior pituitary gland causes an excess secretion of urine, that is, *diuresis*, the identifying characteristic of the disease called *diabetes insipidus*.

outline summary

GENERAL FUNCTIONS

1 Endocrine glands secrete chemicals (hormones) into the blood
2 Hormones perform general functions of communication and control – but a slower, longer-lasting type control than provided by nerve impulses

ANTERIOR PITUITARY GLAND (adenohypophysis)

1 Names of major hormones
   a Thyroid-stimulating hormone (TSH)
   b Adrenocorticotrophic hormone (ACTH)
   c Follicle-stimulating hormone (FSH)
   d Luteinizing hormone (LH)
   e Growth hormone (GH)
   f Prolactin (lactogenic hormone)
2 Functions of major hormones
   a TSH – stimulates growth of thyroid gland; also stimulates it to secrete thyroid hormone
   b ACTH – stimulates growth of adrenal cortex and stimulates it to secrete glucocorticoids (mainly cortisol)
   c FSH – initiates growth of graafian follicles each month in ovary and stimulates one or more follicles to develop to stage of maturity and ovulation; FSH also stimulates estrogen secretion by developing follicles
   d LH – acts with FSH to stimulate estrogen secretion and follicle growth to maturity; LH causes ovulation, is "the ovulating hormone"; LH causes luteinization of the ruptured follicle and stimulates progesterone secretion by corpus luteum
   e GH – stimulates growth by accelerating protein anabolism; also, tends to accelerate fat catabolism and slow glucose catabolism; by slowing glucose catabolism, GH tends to increase blood sugar to higher than normal level (hyperglycemia)
   f Prolactin (lactogenic hormone) – stimulates breast development during pregnancy and secretion of milk after delivery of baby

POSTERIOR PITUITARY GLAND (neurohypophysis)

1 Names of hormones
   a Antidiuretic hormone (ADH)
   b Oxytocin
2 Functions of hormones
   a ADH – accelerates water reabsorption from urine in kidney tubules into blood, thereby decreasing urine secretion (antidiuresis)
   b Oxytocin – stimulates pregnant uterus to contract; may initiate labor; causes glandular cells of breast to release milk into ducts

THYROID GLAND

1 Names of hormones
   a Thyroid hormone
   b Calcitonin (thyrocalcitonin)
2 Functions of hormones
   a Thyroid hormone – accelerates catabolism (tends to increase basal metabolic rate)
   b Calcitonin – tends to decrease blood calcium concentration, perhaps by inhibiting breakdown of bone with release of calcium into blood

PARATHYROID GLANDS

1 Name of hormone – parathormone
2 Function of hormone – parathormone tends to increase blood calcium concentration by increasing breakdown of bone with release of calcium into blood

ADRENAL CORTEX

1 Names of hormones
   a Glucocorticoids (GC's) – chiefly cortisol (hydrocortisone)
   b Mineralocorticoids (MC's) – chiefly aldosterone
   c Sex hormones – small amounts of male hormone (androgens) and female hormone (estrogens) secreted by adrenal cortex of both sexes
2 Functions of hormones
   a Glucocorticoids – enable cells to carry on normal metabolism; high blood concentrations of

glucocorticoids accelerate protein and fat catabolism in liver cells to form glucose (gluconeogenesis), thereby tend to increase blood glucose concentration; increased glucocorticoid secretion is one of first responses body makes to stress; higher than normal concentration of glucocorticoids in blood brings about many other stress responses

b Mineralocorticoids – tend to increase blood sodium and decrease blood potassium concentrations by accelerating kidney tubule reabsorption of sodium and excretion of potassium

ADRENAL MEDULLA

1 Names of hormones – epinephrine (adrenalin) and norepinephrine

2 Functions of hormones – epinephrine and norepinephrine help body resist stress by intensifying and prolonging effects of sympathetic stimulation; increased epinephrine secretion first endocrine response to stress

ISLANDS OF LANGERHANS

1 Names of hormones
 a Glucagon – secreted by alpha cells
 b Insulin – secreted by beta cells

2 Functions of hormones
 a Glucagon tends to increase blood sugar by accelerating liver glycogenolysis (conversion of glycogen to glucose)
 b Insulin tends to decrease blood sugar by accelerating movement of glucose out of blood into cells, which tends to increase glucose metabolism by cells

THYMUS GLAND

Postulated to form lymphocytes before birth and to secrete a hormone essential for immunity

PINEAL BODY

Postulated by some investigators to secrete a hormone called melatonin that may help regulate the functioning of the ovaries

FEMALE SEX GLANDS

Ovaries contain two kinds of cells that secrete hormones – cells of graafian follicles and of corpus luteum; see Chapter 10

MALE SEX GLANDS

Interstitial cells of testes secrete male hormone testosterone that promotes development of male sex organs and of male secondary sex characteristics

new words

antidiuresis
catecholamines
diuresis
gluconeogenesis
glycogenolysis
hyperglycemia
hypoglycemia
luteinization
stress
tetany

review questions

1 What endocrine glands are located in the following parts of the body?
 cranial cavity
 neck
 abdominal cavity
 pelvic cavity

2 What endocrine gland is known as the "master gland"? Why?

3 What endocrine gland secretes each of the following:
 ACTH
 aldosterone
 insulin
 growth hormone
 progesterone

4 What hormone prepares the body for strenuous activity – for "fight or flight," in other words?

5 Many changes occur in the body when it is in a condition of stress; for example, after a person has had major surgery. Name two endocrine glands that greatly increase their secretion of hormones in times of stress. Name the hormones that they secrete.

6 Metabolism changes when the body is in a condition of stress. How does the metabolism of proteins, of fats, and of carbohydrates change, and what hormones cause the changes?

7 What hormone, if a high concentration of it is present in the blood, tends to make us less immune to infectious diseases?

8 What hormone is called the "water-retaining hormone" because it decreases the amount of urine formed?

9 What hormone, if a high concentration of it is present in the blood, tends to make wounds heal more slowly?

10 What hormone is called the "salt-retaining hormone" because it makes the kidneys reabsorb sodium into the blood more rapidly, so that less sodium is lost in the urine?

11 Name two hormones or more that tend to increase blood sugar.

12 What hormone speeds up the rate of catabolism – that is, makes you burn up your foods faster?

13 What is the main function of each of the following:
 ACTH
 ADH
 epinephrine
 insulin
 parathyroid hormone
 thyroid hormone
 testosterone

14 Which hormone is called the ovulating hormone?

15 Does acromegaly result from deficient or excess secretion of what hormone?

16 Does cretinism result from deficient or excess secretion of what hormone?

17 Does Cushing's syndrome result from deficient or excess secretion of what hormone?

18 Does diabetes insipidus result from deficient or excess secretion of ADH or insulin?

19 Does diabetes mellitus result from deficient or excess secretion of ADH or insulin?

20 What endocrine disorder produces giantism?

unit four

Systems that process and distribute foods and eliminate wastes

6

The digestive system

You may already know quite a few facts about the digestive system. You probably know the names of the digestive organs, that some of them form a tube that extends through the body all the way from the mouth to the rectum, and that mucous membrane lines this tube. You most likely know what functions the digestive system performs—that it digests foods, absorbs the digested foods, and eliminates the wastes left over. Figure 1-3, p. 6, lists the names of the organs that compose the digestive system. Figure 6-1, p. 91, shows their locations.

This chapter relates information about the following: the peritoneum, the structure and functions of the various digestive organs, and the complex processes of digestion, absorption, and metabolism.

Peritoneum

The peritoneum is a thin, moist, slippery membrane. It consists of parietal and visceral layers. The *parietal peritoneum* lines the abdominal cavity, whereas the *visceral peritoneum* forms a covering that adheres to the surface of each abdominal organ. The small space between the parietal peritoneum and visceral peritoneum is called the *peritoneal space*. It contains just enough peritoneal fluid to keep both layers of the membrane moist so they can slide freely and easily against each other during breathing and digestive movements. In some diseases an excess of fluid accumulates in the peritoneal space resulting in a condition called *ascites*. You may have heard it called "dropsy."

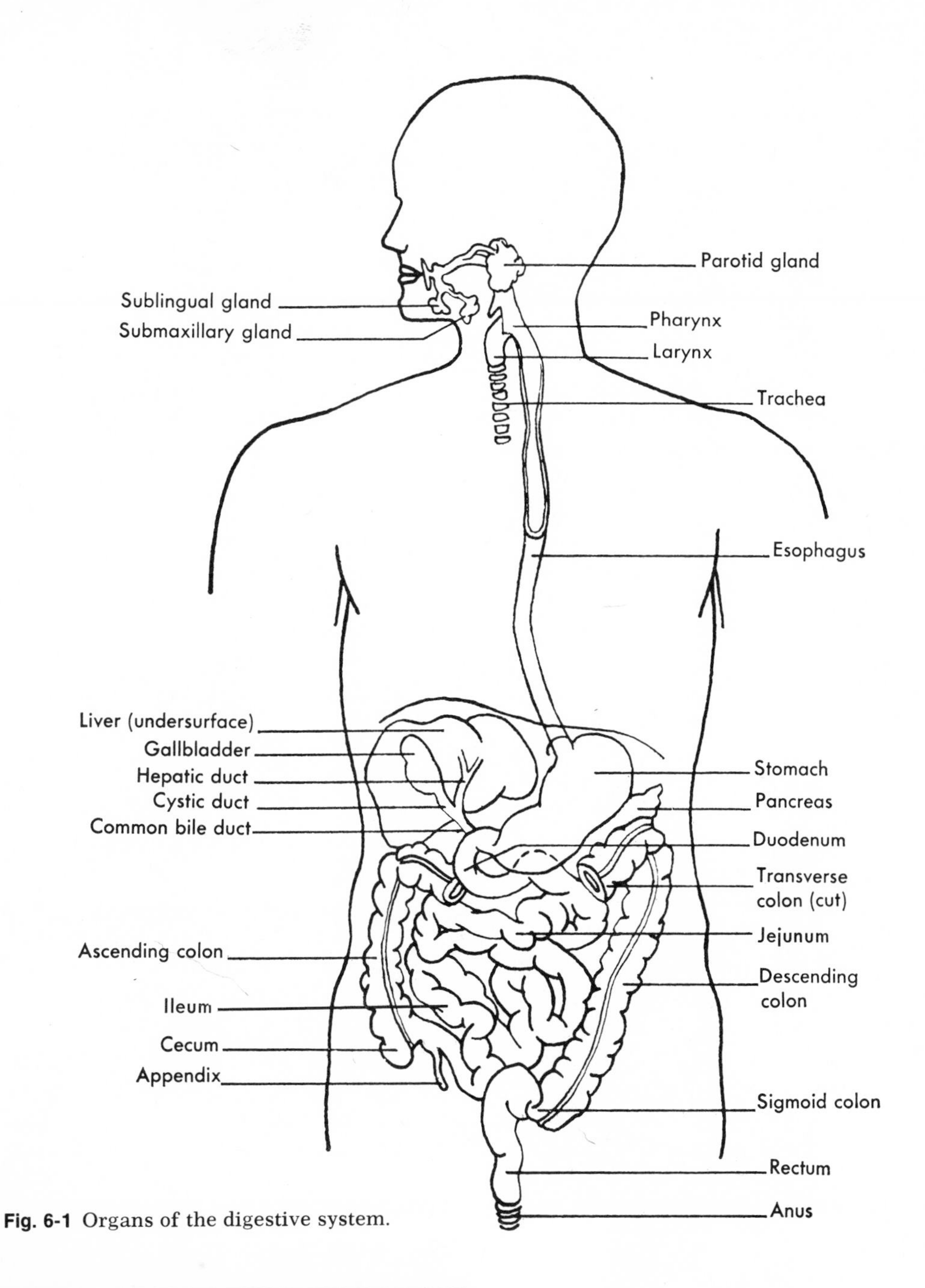

Fig. 6-1 Organs of the digestive system.

Stomach

The stomach is a kind of pouch that food enters after it has been chewed, swallowed, and passed through the esophagus. The stomach looks small after it is emptied, not much bigger than a large sausage, but it expands considerably after a large meal. Have you ever felt so uncomfortably full after eating that you could not take a deep breath? If so, it probably meant that your stomach was so full of food that it occupied more space than usual and pushed up against the diaphragm. This made it hard for the diaphragm to contract and move downward as much as necessary for a deep breath.

Three layers of smooth muscle, with fibers running lengthwise, around, and obliquely in the stomach wall, make the stomach one of the strongest internal organs—well able to break up food into tiny particles and to mix them thoroughly with the gastric juice. Stomach muscle contractions take part in producing *peristalsis,* the movement that propels food down the length of the digestive tract. Mucous membrane lines the stomach; it contains thousands of microscopic glands that secrete gastric juice and hydrochloric acid into the stomach. When the stomach is empty, its lining lies in folds called *rugae.*

The lower part of the stomach is called the *pylorus.* It is a narrow section that joins the first part of the small intestine (the *duodenum*). Food is held in the stomach by the pyloric sphincter muscle long enough for partial digestion. The sphincter consists of circular smooth muscle fibers that stay contracted most of the time and thereby close off the opening of the pylorus into the duodenum. The fibers relax at intervals when part of the food is ready to leave the stomach, but sometimes they go into a spasm and do not relax normally; this condition is referred to as *pylorospasm.* It occurs fairly often in babies.

Small intestine

The small intestine seems to be misnamed if we notice only its length—it is about 20 feet long; however, it is noticeably smaller around than the large intestine so that in this respect its name is appropriate. Different names identify different sections of the small intestine. In the order in which food passes through them, they are the *duodenum, jejunum,* and *ileum.* Find each of these parts of the small intestine in Figure 6-1.

The mucous lining of the small intestine, like that of the stomach, contains thousands of microscopic glands. They are called *intestinal glands,* and they secrete the intestinal digestive juice. We should mention, also, something about the structure of the lining of the small intestine that makes it especially well suited to its function of food and water absorption. It is not perfectly smooth as it appears to the naked eye. Instead it has thousands of tiny "fingers," called *villi.* Under the microscope these can be seen projecting into the hollow interior of the intestine. Inside each of these villi lies a rich network of blood and lymph capillaries. Turn to p. 93 and look at Figure 6-2. Hundreds and hundreds of villi jut inward from the mucous lining. Imagine the lining as perfectly smooth without any villi; think how much less surface area there would be for contact between capillaries and intestinal lining. Consider what an advantage a large contact area offers for faster absorption of food from the intestine into the blood and lymph—one more illustration of the principle that structure determines function.

Smooth muscle in the wall of the small intestine contracts to produce peristalsis, the wormlike movements that move food through the tract.

Large intestine

The large intestine forms the lower part of the digestive tract. Its divisions—cecum, ascending colon, transverse colon, descending colon, sigmoid colon, and rectum—are shown in Figure 6-1. The rectum is about 7 or 8 inches long. Its external opening is called the *anus.* Two sphincter mus-

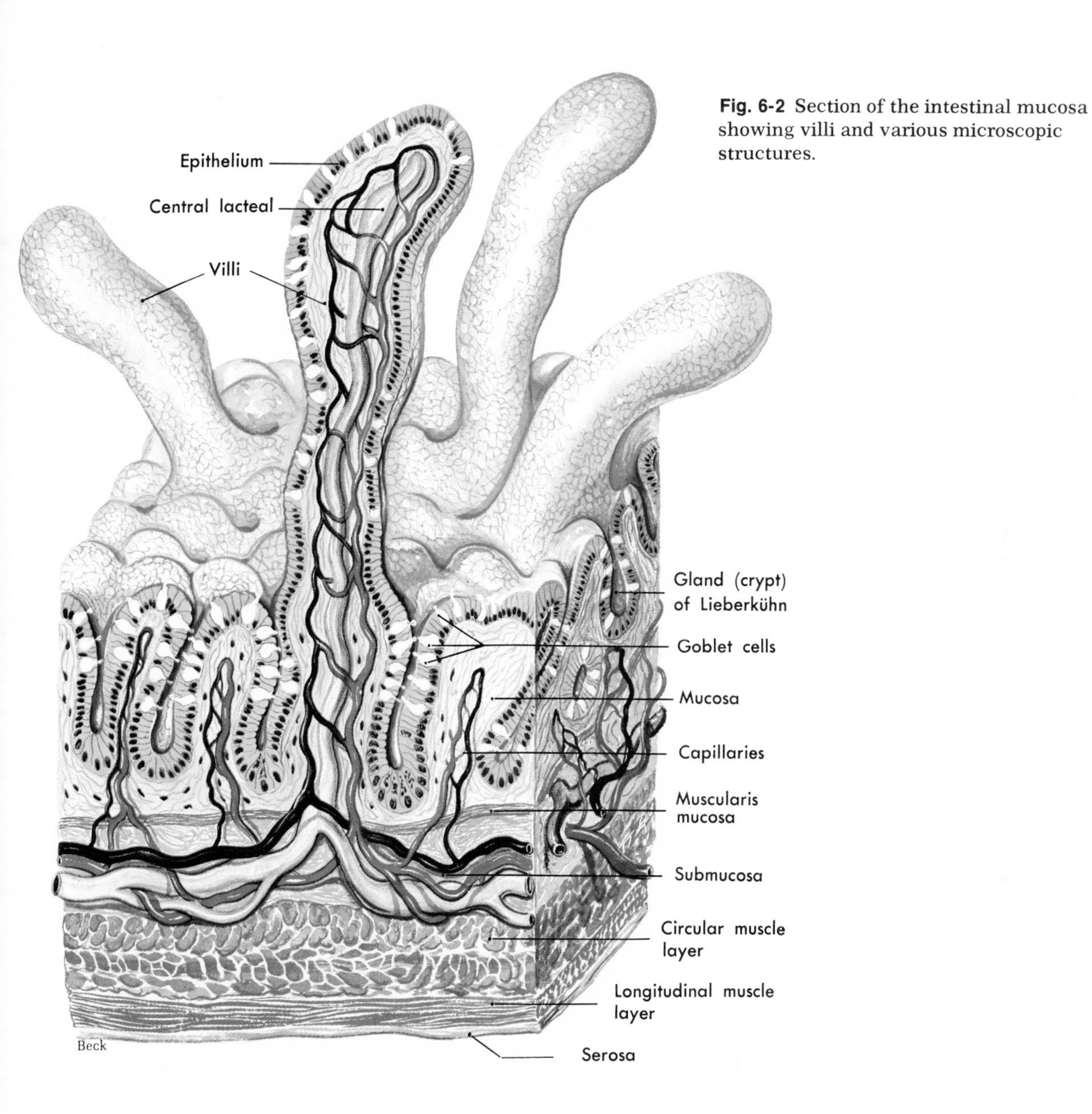

Fig. 6-2 Section of the intestinal mucosa showing villi and various microscopic structures.

cles stay contracted to keep the anus closed except during defecation. Smooth, or involuntary, muscle composes the inner anal sphincter, but striated, or voluntary, muscle composes the outer one. This anatomical fact sometimes becomes highly important from a practical standpoint. For example, often after a person has had a stroke, the voluntary anal sphincter becomes paralyzed. This means, of course, that the individual will have no control over bowel movements. Or, in hospital language, the patient has "involuntary defecations."

Accessory digestive organs

By accessory digestive organs we mean organs that are associated with the digestive tract but that do not form part of the tract itself. Incidentally, the terms "alimentary canal" and "gastrointestinal tract" (or GI tract for short) mean the same thing as the term "digestive tract"—namely, the mouth, pharynx, esophagus, stomach, and intestines. Accessory digestive organs, on the other hand, are the teeth, tongue, salivary glands, pancreas, liver, gallbladder, and vermiform appendix.

Teeth

By the time a baby is 2 years old, he probably has his full set of 20 baby teeth. By the time a young adult is somewhere between 17 and 24 years old, he usually has his full set of 32 permanent teeth. The average age for cutting the first tooth is about 6 months, and the average age for losing the first baby tooth and starting to cut the permanent teeth is about 6 years. Figure 6-3 gives the names of the teeth and shows which ones are lacking in the deciduous or baby set. Figure 6-4 illustrates tooth structure.

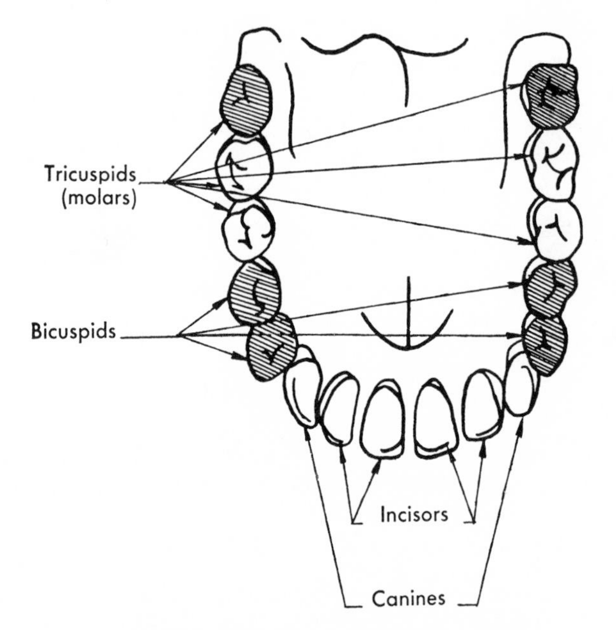

Fig. 6-3 Teeth. The permanent set of teeth includes all of those shown. (Both the upper and lower jaws have the same number and arrangement of teeth.) The deciduous set or "baby teeth" lacks the teeth striped in the diagram.

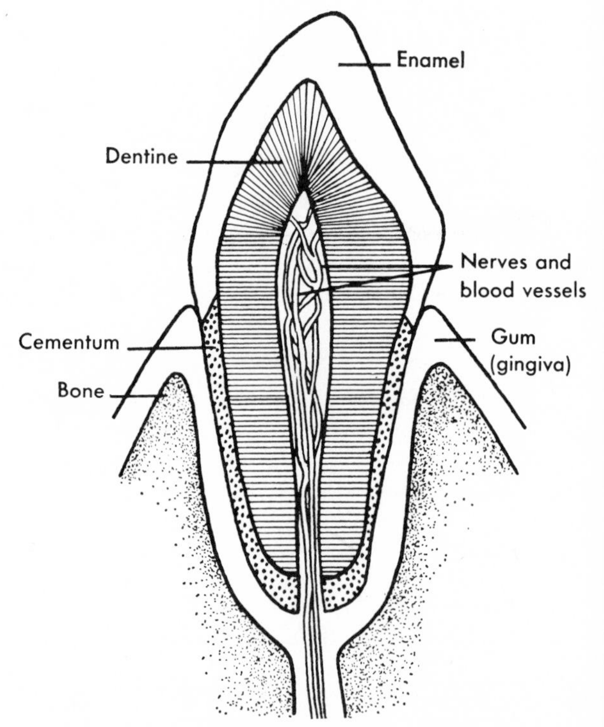

Fig. 6-4 Longitudinal section of an incisor tooth.

Salivary glands

The *parotid glands* lie just below and in front of each ear at the angle of the jaw—an interesting anatomical fact because it explains why people who have mumps (infection of the parotid gland) often complain that it hurts when they open their mouths or chew; these movements squeeze the tender inflamed gland. To see the openings of the ducts of the parotid glands, look on the insides of your cheeks opposite the second molar tooth, on either side of the upper jaw. Besides the parotid glands there are also two other pairs of salivary glands—the *submandibular* and the *sublingual glands*. Their ducts open into the floor of the mouth.

Liver and gallbladder

The liver is the largest gland in the body and one of the most important. It fills the upper right section of the abdominal cavity and even extends part way over onto the left side. One of its many functions is to secrete *bile*. This digestive juice drains out of the liver by way of the *hepatic ducts*. Between meals, bile goes up the cystic duct into the gallbladder (located on the undersurface of the liver) for concentration and storage. After meals, when fats enter the duodenum, the gallbladder ejects bile into the cystic duct, down the common bile duct, and into the duodenum. Blood normally contains some bile pigments; however, if it contains too much, the skin looks yellow and we say that the person has *jaundice*. One thing that can cause jaundice is the absorption of excessive amounts of bile into the blood, caused by a blocking off of one or more of the ducts that drain bile out of the liver into the intestine. Examine Figure 6-5. If a gallstone were to block the cystic duct, would this prevent drainage of bile from the liver into the intestine, or would it merely prevent bile from entering or leaving the gallbladder? A gallstone or any other obstruction of either the hepatic ducts or the common bile duct will prevent bile from draining from the liver into the intestine and, therefore, will produce jaundice. Obstruction of the cystic duct, however, will do neither.

In addition to secreting bile, liver cells perform other functions necessary for healthy survival. Two of the most important are these: they help keep the amount of sugar in the blood at a normal level and they make various blood proteins. When blood sugar (mainly glucose) starts to increase, liver cells change some of it to glycogen and store this compound. But when blood sugar starts to decrease, they change the glycogen back to glucose and release it into the blood. Normally, liver cells are busy after meals changing glucose to glycogen. *Glycogenesis* is the name of this process. It prevents blood sugar from increasing to a dangerously high level when it is being absorbed from the intestine into the blood. Later in the between-meal period when glucose absorption has ended, liver cells carry on either or both of two processes to add glucose to the blood—first, glycogenolysis and then, if needed, gluconeogenesis. *Glycogenolysis* consists of a series of chemical reactions by which glycogen stored in liver cells is changed back to glucose and released into the blood. *Gluconeogenesis* literally means "generating new glucose." It consists of a series of chemical reactions by which liver cells make glucose out of proteins or fats. Both glycogenolysis and gluconeogenesis are essential cogs in the body's complex machinery for maintaining homeostasis of blood sugar. They both help prevent a dangerous decrease in blood sugar when glucose is leaving blood for use by body cells but not entering it from the intestine.

Liver cells synthesize several kinds of protein compounds. They release some of them into the blood, where these com-

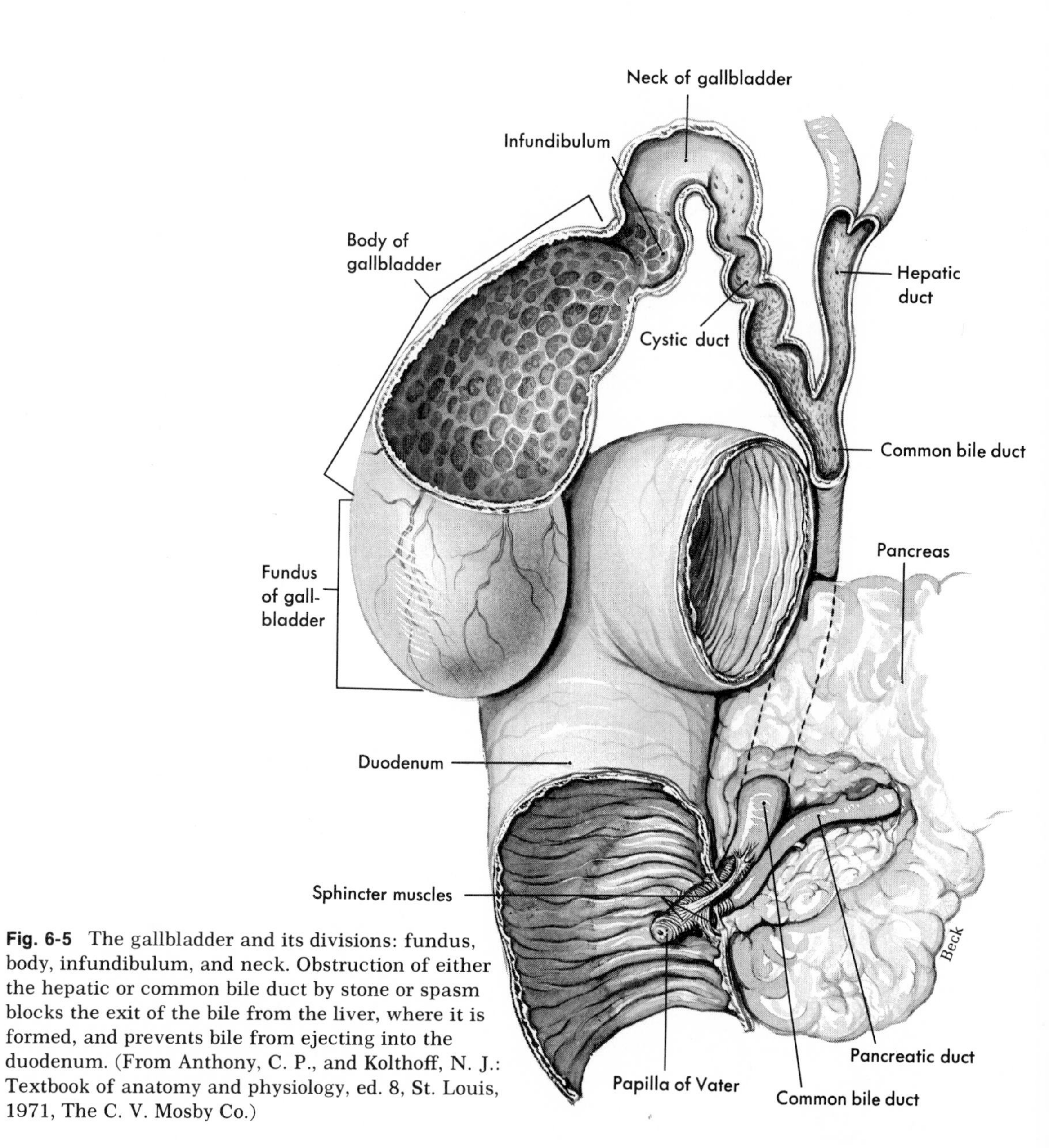

Fig. 6-5 The gallbladder and its divisions: fundus, body, infundibulum, and neck. Obstruction of either the hepatic or common bile duct by stone or spasm blocks the exit of the bile from the liver, where it is formed, and prevents bile from ejecting into the duodenum. (From Anthony, C. P., and Kolthoff, N. J.: Textbook of anatomy and physiology, ed. 8, St. Louis, 1971, The C. V. Mosby Co.)

pounds are called the "blood proteins" or "plasma proteins." Prothrombin and fibrinogen are the names of two of the blood proteins formed by liver cells. What do you remember about the function of these substances? (See p. 118.)

The first steps in protein and fat catabolism consist of chemical reactions carried on by liver cells.

Now, to sum up liver functions for yourself, read over the preceding paragraphs. You have learned that the liver functions in the metabolism of all three kinds of foods. Actually, it plays a major role in metabolism. The liver is, therefore, one of our vital organs. Patients who have extensive liver disease are desperately sick individuals.

Pancreas

The pancreas lies behind the stomach. Some of its cells secrete *pancreatic juice* into ducts. This most important digestive juice drains out of the *pancreatic duct* into the duodenum at the same place that the bile enters. Some other cells located in the pancreas secrete *insulin* into the blood instead of into ducts. In other words, these cells are ductless, or endocrine, glands. Because they are separate from the cells that secrete pancreatic juice, they are called islands—*islands of Langerhans*. Insulin, as discussed on p. 85, is a hormone absolutely vital for maintaining a normal amount of blood sugar.

Appendix

The appendix is a dead-end tube off the cecum. Although it serves no known function, it is often a nuisance when its mucous lining becomes inflamed—the well-known affliction, appendicitis.

Digestion

All of the digestive organs work together to perform the function known as digestion. Digestion is the process that prepares foods for absorption; that is, it changes foods into substances that can pass through the mucous membrane lining of the small intestine into the blood and eventually into cells. Digestion consists of several mechanical processes and many chemical reactions. The mechanical processes—chewing, swallowing, and peristalsis—break the food into tiny particles, mix them with the digestive juices, and move them along the digestive tract. Molecules that compose the foods we eat are too large for absorption. Digestion changes this. Chemical reactions catalyzed by the *digestive enzymes* break down the large nonabsorbable food molecules into smaller molecules that can readily pass through the intestinal mucosa into blood or lymphatic capillaries.

Protein digestion starts in the stomach. Two enzymes (rennin and pepsin) in the gastric juice cause the giant protein molecules to break up into somewhat simpler compounds. Then, in the intestine other enzymes finish the job of protein digestion. Pancreatic juice contains one protein-digesting enzyme (trypsin), and intestinal juice contains another (erepsin). Every protein molecule is made up of many amino acids joined together. When enzymes have split up the large protein molecule into its separate amino acids, protein digestion is completed. Hence, the end product of protein digestion is amino acids. For obvious reasons, amino acids are also referred to as "protein building blocks."

Very little carbohydrate digestion occurs before food reaches the small intestine. (Starches and sugars are carbohydrates.) Gastric juice contains no enzymes that act on carbohydrates, and the enzymes in saliva usually have too little time to do their work because so many of us swallow our food so fast. But once the food reaches the small intestine, pancreatic and intestinal juice enzymes digest the starches and sugars. A pancreatic enzyme (amylopsin or amylase) starts the process by changing starches into maltose. Three intestinal enzymes digest sugars, changing them into glucose (also called dextrose). Maltase digests maltose (malt sugar), sucrase digests sucrose (ordinary cane sugar), and lactase digests lactose (milk sugar). The main end product of carbohydrate digestion is glucose, a simple sugar.

Not only very little carbohydrate digestion but also very little fat digestion occurs before food reaches the small intestine. An enzyme in gastric juice (gastric lipase) digests some fat in the stomach, but most fats go undigested until after bile emulsifies them—that is, breaks the fat droplets into very small droplets. After this takes place, the pancreatic enzyme (steapsin or pancreatic lipase) splits up the fat molecules into fatty acids and glycerol (glycerin). The end products of fat digestion, then, are fatty acids and glycerol. Have you noticed that only one digestive juice—pancreatic juice—can digest all three kinds of foods? In what organ does the pancreatic juice do its work?

Table 6-1 summarizes the main facts about chemical digestion. Enzyme names indicate the type of food digested by the enzyme. For example, the name *amylase* indicates that the enzyme digests carbohydrates (starches and sugars). The name *protease* indicates a protein-digesting enzyme, and the name *lipase* means a fat-digesting enzyme. When carbohydrate digestion has been completed, starches and complex sugars (polysaccharides) have been changed mainly to glucose, a simple sugar (monosaccharide). The end product of protein digestion, on the other hand, is amino acids. Fatty acid and glycerol are the end products of fat digestion. Use information in Table 6-1 to answer questions 10 to 15, p. 103.

Table 6-1. Chemical digestion

Digestive juices and enzymes	*Enzyme digests (or hydrolyzes)*	*Resulting product**
Saliva		
Amylase (ptyalin)	Starch (polysaccharide or complex sugar)	Maltose (a disaccharide or double sugar)
Gastric juice		
Protease (pepsin) plus hydrochloric acid	Proteins, including casein	Proteoses and peptones (partially digested proteins)
Lipase (of little importance)	Emulsified fats (butter, cream, etc.)	*Fatty acids* and *glycerol*
Bile contains no enzymes	Large fat droplets (unemulsified fats)	Small fat droplets or emulsified fats
Pancreatic juice		
Protease (trypsin)†	Proteins (either intact or partially digested)	Proteoses, peptides, and *amino acids*
Lipase (steapsin)	Bile–emulsified fats	*Fatty acids* and *glycerol*
Amylase (amylopsin)	Starch	Maltose
Intestinal juice (succus entericus)		
Peptidases	Peptides	*Amino acids*
Sucrase	Sucrose (cane sugar)	*Glucose* and *fructose*‡ (simple sugars or monosaccharides)
Lactase	Lactose (milk sugar)	*Glucose* and *galactose* (simple sugars)
Maltase	Maltose (malt sugar)	*Glucose* (grape sugar)

*Substances in italics are end products of digestion or, in other words, completely digested foods ready for absorption.
†Secreted in inactive form (trypsinogen); activated by enterokinase, an enzyme in the intestinal juice.
‡Glucose is also called dextrose; fructose is called levulose.

Absorption

When we talk about food being absorbed, we mean that after it is digested it moves through the mucous membrane lining of the small intestine into the blood and lymph. In other words, food absorption means that molecules of amino acids, glucose, fatty acids, and glycerol go from the inside of the intestines into the circulating fluids of the body. Absorption of foods is just as essential a process as the digestion of foods. The reason is fairly obvious. As long as food stays in the intestines, it cannot nourish the millions of cells composing all other parts of the body. Their lives depend upon the absorption of digested food and its transportation to them by the circulating blood. On p. 92 we mentioned something about the structure of the intestinal lining that enables it to absorb foods rapidly. Do you recall what this is?

Metabolism

A good phrase to remember in connection with the word metabolism is "the use of foods," for basically this is what metabolism is—the use the body makes of foods once they have been digested, absorbed, and circulated to cells. It uses them in two ways: as an energy source, and as building blocks for making complex chemical compounds. Before they can be used in these two ways, foods have to enter cells and there undergo many chemical changes. All chemical reactions that release energy from food molecules, together make up the process of catabolism—a vital process since it is the only way that the body has of supplying itself with energy for doing any of its many kinds of work. The many chemical reactions that build food molecules into more complex chemical compounds together constitute the process of anabolism. Catabolism and anabolism together make up the process of metabolism. Reread the discussion of metabolism on pp. 7 to 8. Reexamine Figure 1-5, p. 8.

Something worth noticing is that the amount of food in the blood normally does not change very much, not even when we go without food for many hours, or when we exercise and use a lot of food for energy, or when we sleep and use little food for energy. The amount of glucose, for example, usually stays at about 80 to 120 milligrams in 100 milliliters of blood.

Several hormones help regulate carbohydrate metabolism to keep blood sugar normal. *Insulin* is one of the most important of these. It acts in some way not yet definitely known to make glucose leave the blood and enter the cells at a more rapid rate. As insulin secretion increases, more glucose leaves the blood and enters the cells. The amount of glucose in the blood, therefore, tends to decrease while the rate of glucose metabolism in cells tends to increase. Too little insulin secretion, such as occurs in diabetes mellitus, produces the opposite effects. Less glucose leaves the blood and enters the cells, therefore, more glucose remains in the blood but less glucose is metabolized by cells. In other words, high blood sugar (hyperglycemia) and a low rate of glucose metabolism characterize insulin deficiency. Insulin is the only hormone that functions to lower the blood sugar level. Several other hormones, on the other hand, tend to increase it. Growth hormone secreted by the anterior pituitary gland, hydrocortisone secreted by the adrenal cortex, and epinephrine secreted by the adrenal medulla are three of the most important hormones that tend to increase blood sugar. More information about these hormones and others that help control metabolism appeared on pp. 82 to 85.

The carbohydrate glucose is the body's preferred energy food. Human cells seem to prefer to catabolize glucose rather than other substances, and they do so as long as enough glucose enters them to supply their energy needs. They also anabolize small amounts of carbohydrates. For example, the process of glycogenesis, described on p. 95, is an important kind of carbohydrate anabolism.

Fats, like carbohydrates, are primarily energy foods. If cells have inadequate amounts of glucose to catabolize, they immediately shift to the catabolism of fats for their energy supply. This happens normally whenever a person goes without food for many hours. It happens abnormally in diabetic individuals. Because of an insulin deficiency, too little glucose enters the cells of a diabetic person to supply all of his energy needs, so that the cells catabolize fats to make up the difference. In all persons, fats not needed for catabolism are anabolized and stored in adipose tissue.

Protein catabolism occurs to some extent, but more important is protein anabolism, the process by which the body builds amino acids into complex protein compounds—for example, enzymes and proteins that form the structural parts of the cell.

Metabolic rates

Every nurse, sooner or later, hears the letters BMR. They stand for *basal metabolic rate*. Suppose you were to "have a BMR" test. You would breathe through your mouth into a tube connected to a machine that would measure the amount of oxygen you breathe. You would be lying down but awake. You would not have eaten for 12 hours or more, and the room would be comfortably warm. (The last two sentences describe basal conditions.) From the amount of oxygen you had breathed in, the technician could compute the amount of food your body had catabolized under the basal conditions—could compute your BMR, in other words. The nurse might report this as so many calories (that is, so much heat from catabolized food). More often she reports it simply as "normal" or a certain percentage above or below normal. Your normal BMR might be 1,200 calories.

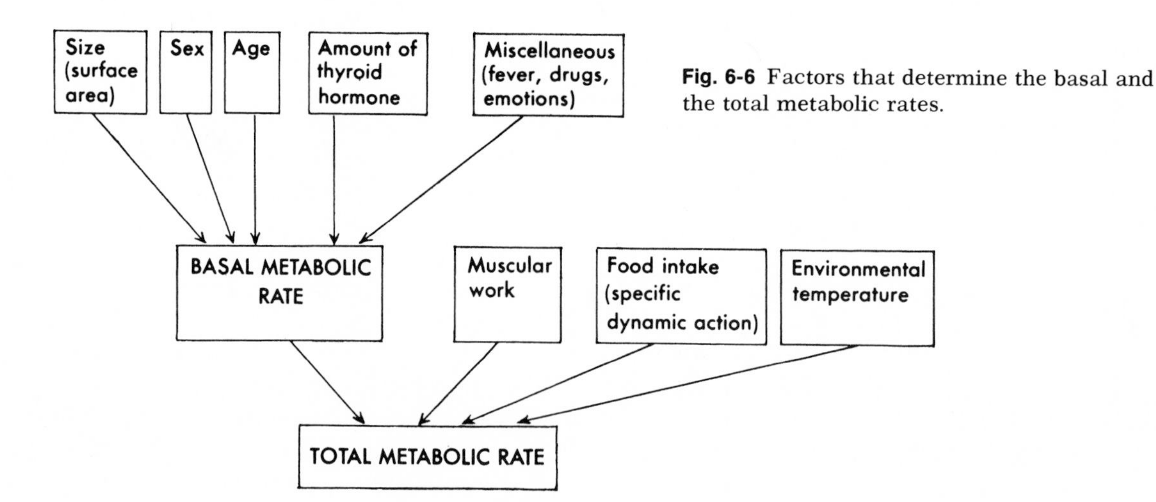

Fig. 6-6 Factors that determine the basal and the total metabolic rates.

This would not be normal for everyone, however. Someone bigger or younger than you might have a normal BMR of 1,400 calories. Someone smaller or someone the same size but older than you, however, might have a normal basal rate of only 1,000 calories. A man normally has a higher basal rate than a woman his same size and age.

A person's BMR represents the amount of food that his body must catabolize each day for him simply to stay alive and awake in a comfortably warm environment. To provide energy for him to do any muscular work and for him to digest and absorb any food, an additional amount of food must be catabolized. How much more food must be catabolized depends upon how much work the individual does. The more active he is, the more food his body must catabolize and the higher his total metabolic rate will be. The *total metabolic rate*, or TMR, is the total amount of energy (expressed in calories) used by the body per day (Figure 6-6).

When a person's TMR equals the total calories in all the food that he eats per day, that person's weight remains constant (except for possible variations resulting from water retention or water loss). When his daily food contains more calories than his TMR, he gains weight; when his daily food contains fewer calories than his TMR, he loses weight. These weight control principles never fail to operate. Nature never forgets to count calories. Reducing diets make use of this knowledge. They contain fewer calories than the TMR of the individual eating the diet.

outline summary

PERITONEUM

Thin, moist membrane consisting of layer that lines abdominal cavity (parietal layer) and layer that adheres to surfaces of abdominal organs (visceral layer); peritoneal fluid in peritoneal space between parietal and visceral layers of peritoneum lubricates peritoneum

STOMACH

1. Size – expands after large meal; about size of large sausage when empty
2. Pylorus – lower part of stomach; pyloric sphincter muscle closes opening of pylorus into duodenum
3. Wall – many smooth muscle fibers; contractions produce churning movements and peristalsis
4. Lining – mucous membrane; many microscopic glands that secrete gastric juice and hydrochloric acid into stomach; mucous membrane lies in folds (rugae) when stomach is empty

SMALL INTESTINE

1. Size – about 20 feet long but only an inch or so in diameter
2. Divisions
   - a Duodenum
   - b Jejunum
   - c Ileum
3. Wall – contains involuntary muscle fibers that contract to produce peristalsis
4. Lining – mucous membrane; many microscopic glands (intestinal glands) secrete intestinal juice; villi (microscopic finger-shaped projections from surface of mucosa into intestinal cavity) contain blood and lymph capillaries

LARGE INTESTINE

1. Divisions
   - a Cecum
   - b Colon – ascending, transverse, descending, and sigmoid
   - c Rectum
2. Opening to exterior – anus
3. Wall – contains smooth muscle fibers that contract to produce churning, peristalsis, and defecation
4. Lining – mucous membrane

ACCESSORY DIGESTIVE ORGANS

Teeth

1. Twenty teeth in temporary set; average age for cutting first tooth about 6 months; set complete at about 2 years of age
2. Thirty-two teeth in permanent set; 6 years about average age for starting to cut first permanent tooth; set complete usually between ages of 17 to 24 years

3 Names of teeth – see Figure 6-3
4 Structures of tooth – see Figure 6-4

Salivary glands

1 Parotid
2 Submaxillary
3 Sublingual

Liver and gallbladder

1 Size and location – largest gland; fills upper right section of abdominal cavity and extends over into left side
2 Functions – secretes bile; helps maintain normal blood sugar by carrying on processes of glycogenesis and glycogenolysis; forms prothrombin, fibrinogen, and certain other blood proteins; also performs several other functions
3 Ducts
   a Hepatic – drains bile from liver
   b Cystic – duct by which bile enters and leaves gallbladder
   c Common bile – formed by union of hepatic and cystic ducts; drains bile from hepatic or cystic ducts into duodenum

Pancreas

1 Location – behind stomach
2 Functions
   a Pancreatic cells secrete pancreatic juice into pancreatic ducts; main duct empties into duodenum
   b Islands of Langerhans – cells not connected with pancreatic ducts; secrete insulin into blood

Appendix

Blind tube off cecum; no known function

DIGESTION

Meaning – changing foods so that they can be absorbed and used by cells

Mechanical digestion

Chewing, swallowing, and peristalsis break food into tiny particles, mix them well with the digestive juices, and move them along the digestive tract

Chemical digestion

Breaks up large food molecules into compounds having smaller molecules; brought about by digestive enzymes

Protein digestion

Starts in stomach; completed in small intestine
1 Gastric juice enzymes, rennin and pepsin, partially digest proteins
2 Pancreatic enzyme, trypsin, completes digestion of proteins to amino acids
3 Intestinal enzyme, erepsin, completes digestion of partially digested proteins to amino acids

Carbohydrate digestion

Mainly in small intestine
1 Pancreatic enzyme, amylopsin, changes starches to maltose
2 Intestinal juice enzymes
   a Maltase changes maltose to glucose
   b Sucrase changes sucrose to glucose
   c Lactase changes lactose to glucose

Fat digestion

1 Gastric lipase changes small amount of fat to fatty acids and glycerin in stomach
2 Bile contains no enzymes but emulsifies fats (breaks fat droplets into very small droplets)
3 Pancreatic lipase changes emulsified fats to fatty acids and glycerin in small intestine

ABSORPTION

1 Meaning – digested food moves from intestine into blood or lymph
2 Where absorption occurs – foods and water from small intestine; water also absorbed from large intestine

METABOLISM

1 Meaning – use of foods by body cells for energy and building complex compounds
2 Catabolism – breaks food molecules down into carbon dioxide and water, releasing their stored energy; oxygen used up in catabolism
3 Anabolism – builds food molecules into complex substances
4 Carbohydrates primarily catabolized for energy but small amounts are anabolized by glycogenesis (a series of chemical reactions that changes glucose to glycogen – occurs mainly in liver cells where glycogen is stored); glycogenolysis is process (series of chemical reactions) by which glycogen is changed back to glucose
5 Blood sugar (or blood glucose) – normally stays between about 80 and 120 milligrams per 100 milliliters of blood; *insulin* accelerates movement of glucose out of blood into cells, therefore tends to decrease blood glucose and increase glucose catabolism
6 Fats both catabolized to yield energy and anabolized to form adipose tissue
7 Proteins primarily anabolized and secondarily catabolized

Metabolic rates

1 Basal metabolic rate (BMR) – rate of metabolism when person is lying down, but awake, when about 12 hours have passed since last meal, and when environment is comfortably warm
2 Total metabolic rate (TMR) – the total amount of energy, expressed in calories, used by the body per day

new words

absorption
ascites
digestion
glycogenesis
glycogenolysis
gluconeogenesis
peristalsis
peritoneum

review questions

1 What organs form the gastrointestinal tract?
2 Identify each of the following structures: jejunum, cecum, colon, duodenum, and ileum.
3 If you inserted 9 inches of an enema tube through the anus, the tip of the tube would probably be in what structure?
4 How many teeth should an adult have?
5 How many teeth should a child $2^1/_2$ years old have? Would he have some of each of the following teeth: incisors, canines, bicuspids, and tricuspids?
6 Identify each of the following:
   rugae
   pylorus
   villi
   islands of Langerhans
   parotid glands
7 What process do liver cells carry on that prevents the level of blood sugar from becoming dangerously high following a heavy meal?
8 What two processes do liver cells carry on that prevent a dangerously low blood sugar from developing during fasting and between meals?
9 In what organ does the digestion of starches begin?
10 What digestive juice contains no enzymes?
11 Only one digestive juice contains enzymes for digesting all three kinds of food. Which juice is this?
12 Which kind of food is not digested in the stomach?
13 Which digestive juice emulsifies fats?
14 What three digestive juices act on foods in the small intestine?
15 What juices digest carbohydrates? proteins? fats?
16 Explain as briefly and clearly as you can what each of the following terms means:
   digestion
   absorption
   metabolism
   anabolism
   catabolism
17 Explain why you think the following statement is true or false: "If you do not want to gain or lose weight but just stay the same, you must eat just enough food to supply the calories of your BMR. If you eat more than this, you will gain; if you eat less than this, you will lose."

7

The respiratory system

No one needs to be told how important his respiratory system is. Everyone knows this intuitively. For example, think how panicky you would feel if suddenly you could not breathe for a few seconds. Of all the substances that cells and therefore the body as a whole must have to survive, oxygen is by far the most crucial. A person can live a few weeks without food, a few days without water, but only a few minutes without oxygen. The respiratory system serves the body much as a lifeline to an oxygen tank serves a deep-sea diver. In this chapter the structural plan of the respiratory system will be considered first, then the respiratory organs will be discussed individually, and finally some facts about respiration that are important for a nurse to know will be given.

Structural plan

The respiratory organs are the nose, pharynx, larynx, trachea, bronchi, and lungs. Their basic plan is that of a tube with many branches ending in millions of extremely tiny, extremely thin-walled sacs called *alveoli.* A network of capillaries fits like a tight-fitting hairnet around each microscopic alveolus. Incidentally, this is a good place for us to think again about a principle already mentioned several times, namely, that structure determines function. The function of alveoli—in fact, the function of the entire respiratory system—is to bring air close enough to blood for oxygen to get into the blood and carbon dioxide to get out of it. Two facts about the structure of alveoli make them able to

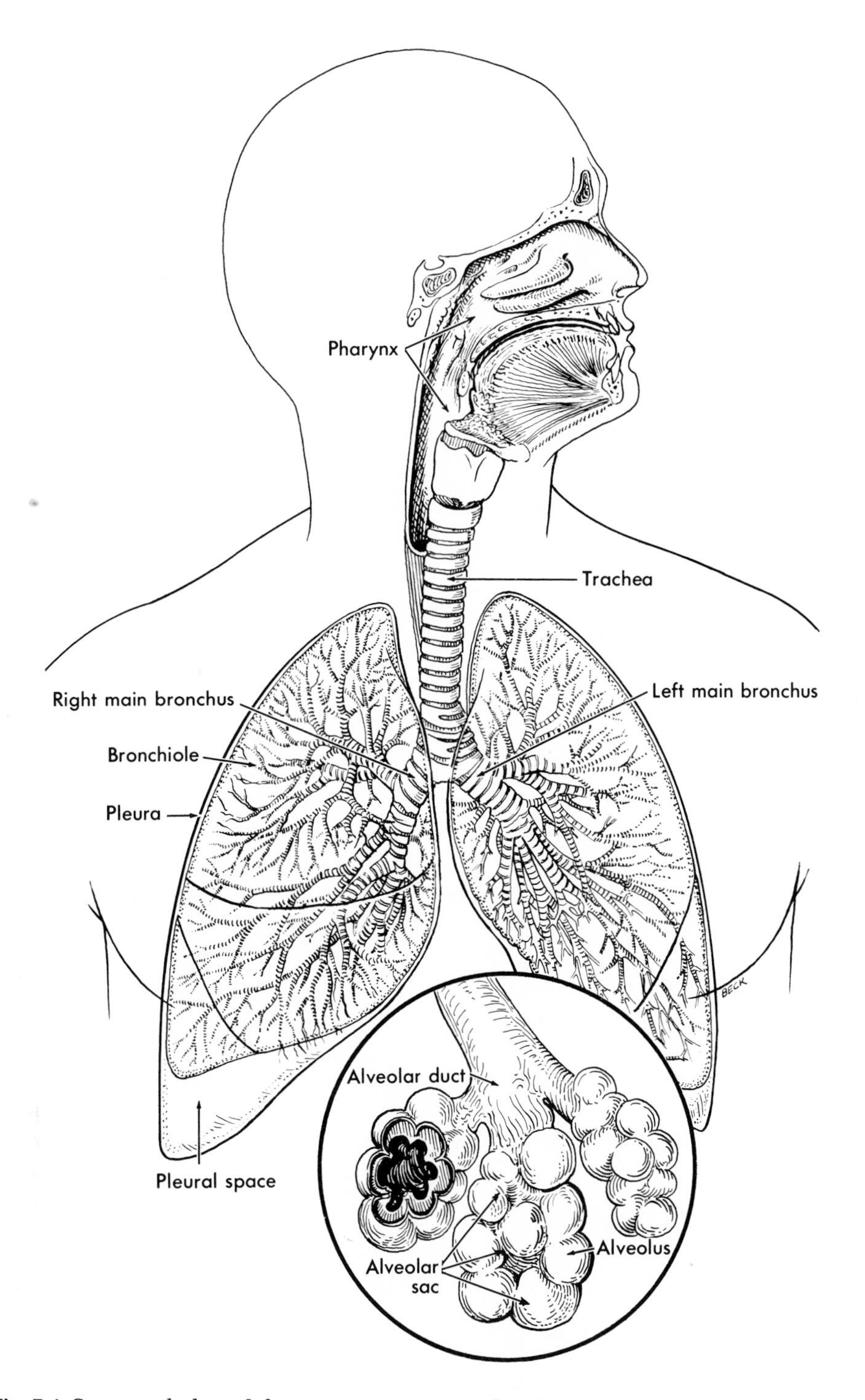

Fig. 7-1 Structural plan of the respiratory organs showing the pharynx, trachea, bronchi, and lungs. The inset shows the grapelike alveolar sacs where the interchange of oxygen and carbon dioxide takes place through the thin walls of the alveoli. Capillaries (not shown) surround the alveoli.

perform this function admirably. First, the wall of each alveolus is made up of a single layer of cells and so are the walls of the capillaries around it. This means that between the blood in the capillaries and the air in the alveolus there is a barrier probably less than 1/5,000 of an inch thick! Second, there are millions of alveoli. This means that together they make an enormous surface (in the neighborhood of 1,100 square feet, an area many times larger than the surface of the entire body) where large amounts of oxygen and carbon dioxide can rapidly be exchanged.

Nose

Air enters the respiratory tube through the nostrils (nares) into the right and left nasal cavities. A partition, the nasal septum, separates these two cavities; mucous membrane lines them. The surface of the nasal cavities is moist from mucus and warm from blood flowing just under it. Nerve endings responsible for the sense of smell are located in the nasal mucosa. Four pairs of sinuses (frontal, maxillary, sphenoidal and ethmoidal) drain into the nasal cavities. The nasal mucosa also lines the sinuses, and since it becomes infected whenever we have a cold, sinus infection often develops from colds.

Pharynx

The pharynx is what most of us call the throat. It serves the same purpose for the respiratory and digestive tracts as a hallway serves for a house. Both air and food pass through the pharynx on their way to the lungs and the stomach, respectively. Air enters the pharynx from the two nasal cavities and leaves it by way of the larynx; food enters it from the mouth and leaves it by way of the esophagus. Also, the right and left eustachian (auditory) tubes open into the pharynx; they connect each middle ear with the pharynx. Two pairs of organs that seem to give more trouble than service to the body (namely, the *tonsils* and the *adenoids*) are also located in the pharynx.

Larynx

The larynx, or voice box, is located just below the pharynx. It is composed of several pieces of cartilage. You know the largest of these (the *thyroid cartilage*) as the "Adam's apple." One of the cartilages of the larynx, the epiglottis, acts as a lid to close off the larynx when we swallow. Occasionally this lid does not work properly; then we cough and choke and say that we have "swallowed down our Sunday throat," which means that food or liquid has entered the larynx where only air should go.

Two short fibrous bands called the vocal cords stretch across the interior of the larynx. Muscles that attach to the cartilages of the larynx can pull on these cords in such a way that they become either tense and short or long and relaxed. When they are tense and short, the voice sounds high pitched; when they are long and relaxed, it sounds low pitched.

Trachea

By pushing with your fingers against your throat about an inch above the breast bone, you can feel the shape of the trachea, or windpipe. Only if you use considerable force can you squeeze it closed. Nature has taken precautions to keep this lifeline open. Its framework is made of an almost noncollapsible material—namely, 15 or 20 C-shaped rings of cartilage placed one above the other with only a little soft tissue between them. Despite this structural safeguard, closing of the trachea does sometimes occur. A tumor or an infection may enlarge the lymph nodes of the neck so much that they squeeze the trachea shut, or a person may aspirate (breathe in) a piece of food or something else that blocks the windpipe. Since air has no other way to get to the lungs, complete tracheal obstruction causes death in a matter of minutes.

Bronchi, bronchioles, and alveoli

One way to picture the thousands of air tubes that make up the lungs is to think of an upside-down tree. The trachea is the main trunk of this tree; the right bronchus (the tube leading into the right lung) and the left bronchus (the tube leading into the left lung) are the trachea's first branches. In each lung they branch into smaller bronchi, which branch into bronchioles. The smallest bronchioles end in structures shaped like miniature bunches of grapes (Figure 7-1). The smallest bronchioles subdivide into microscopic-sized tubes called *alveolar ducts,* which resemble the main stem of a bunch of grapes. Each alveolar duct ends in several *alveolar sacs,* each of which resembles a cluster of grapes, and the wall of each alveolar sac is made up of numerous *alveoli,* each of which resembles a single grape. How well the structure of the alveoli suits them to their function was discussed on pp. 104 and 106.

Because the air tubes of the bronchial tree are imbedded in connective tissue, you cannot see them by looking at the lungs from the outside. But if you cut into the lungs, the bronchi and bronchioles show up clearly.

Lungs and pleura

The lungs are fairly large organs. They fill the entire chest cavity (all but the space in the center occupied mainly by the heart and large blood vessels). The narrow part of each lung, up under the collarbone, is its *apex;* the broad lower part, resting on the diaphragm, is its *base.* The *pleura* covers the outer surface of the lungs and lines the inner surface of the rib cage. The pleura resembles the peritoneum in structure and function, but differs from it in location. Both are extensive, thin, moist, slippery membranes. Both line a large, closed cavity of the body and cover organs located in them. The *parietal pleura* lines the thoracic cavity, the *visceral pleura* covers the lungs, and the *pleural space* lies between them. Where are the parietal peritoneum, the visceral peritoneum, and the peritoneal space located? (Chapter 6 will tell you, if you do not know.)

Normally, the pleural space contains just enough fluid to make both portions of the pleura moist and slippery and able to glide easily against each other as the lungs expand and deflate with each breath. However, the pleural space sometimes becomes distended with a large amount of fluid. The extra fluid presses on the lungs and makes it hard for the patient to breathe. In this case, the physician may decide to remove the excess pleural fluid by means of a hollow, tubelike instrument that he pushes through the patient's chest wall into the pleural space. If you were to ask the patient what the doctor did to him, he probably would answer, "He tapped my chest." If you asked the doctor, he might answer with the technical name of the procedure. He might say that he had done a *thoracentesis* on the patient.

Pleurisy and pneumothorax are other terms having to do with the pleura and pleural space. *Pleurisy* (or pleuritis) is an inflammation of the pleura. *Pneumothorax* is the injection of air into the pleural space of one side of the chest. The additional air presses on the lung on that side and collapses it. While it remains collapsed, the lung does not function in breathing. Pneumothorax, therefore, is a procedure performed to rest a diseased lung.

Respiration

Respiration means exchange of gases (oxygen and carbon dioxide) between a living organism and its environment. If the organism consists of only one cell, gases can move directly between it and the environment. If, however, the organism consists of billions of cells, as do our bodies, then most of its cells are too far from the air for a direct exchange of gases. To overcome this difficulty, an organ – the lungs – is provided where air and a circulating fluid (blood) can come close enough to each other for oxygen to move out of the air into blood while carbon dioxide moves out of the blood into air. This exchange of gases between air and blood is what we call breathing, or respiration. The movement of gases between the blood and cells, however, is also respiration – cellular respiration, or cell breathing.

Exchange of gases in the lungs

As blood flows through the thousands of tiny lung capillaries, carbon dioxide leaves it and oxygen enters it. This two-way exchange of gases between the blood in the lung capillaries and the air in the alveoli comes about in the following way. Venous blood enters lung capillaries. (These are the only capillaries where this is true; arterial blood enters all other capillaries – "tissue capillaries" as they are called.) The hemoglobin inside many of the red blood cells in venous blood is combined, as you can see in Figure 7-2, with carbon dioxide rather than with oxygen, and so is called carbaminohemoglobin. As the blood flows along through the lung capillaries, carbaminohemoglobin breaks down into carbon dioxide and hemoglobin. The hemoglobin then picks up oxygen molecules while carbon dioxide molecules move out of the blood. They pass quickly through the thin capillary-alveolar membrane into the alveoli. Then from the alveoli, carbon dioxide leaves the body in the expired air.

Some carbon dixoide is also present in blood in the form of carbonic acid, formed by water combining with carbon dioxide as shown by the following equation:

$$H_2O + CO_2 \rightarrow \underset{\text{(carbonic acid)}}{H_2CO_3}$$

Some of the carbon dioxide that leaves the lung capillary blood has come from the reversing of the above equation, from carbonic acid breaking down again to become water and carbon dioxide. Therefore, as carbon dixoide leaves the blood in the lung capillaries, both the carbonic acid content of blood and its carbon dioxide content decrease.

The exchange of gases between lung capillary blood and the alveolar air (carbon dioxide out of the blood, oxygen into the blood) changes venous blood to arterial blood. Figure 7-2 shows these changes. In summary, the exchange of gases in the lung capillaries causes arterial blood to differ from venous blood in these ways: arterial blood contains more oxygen, but less carbon dioxide and less carbonic acid than does venous blood. Also arterial blood is more alkaline than venous blood, because arterial blood contains less carbonic acid.

Imagine a patient who hyperventilates over a period of hours. (Hyperventilation

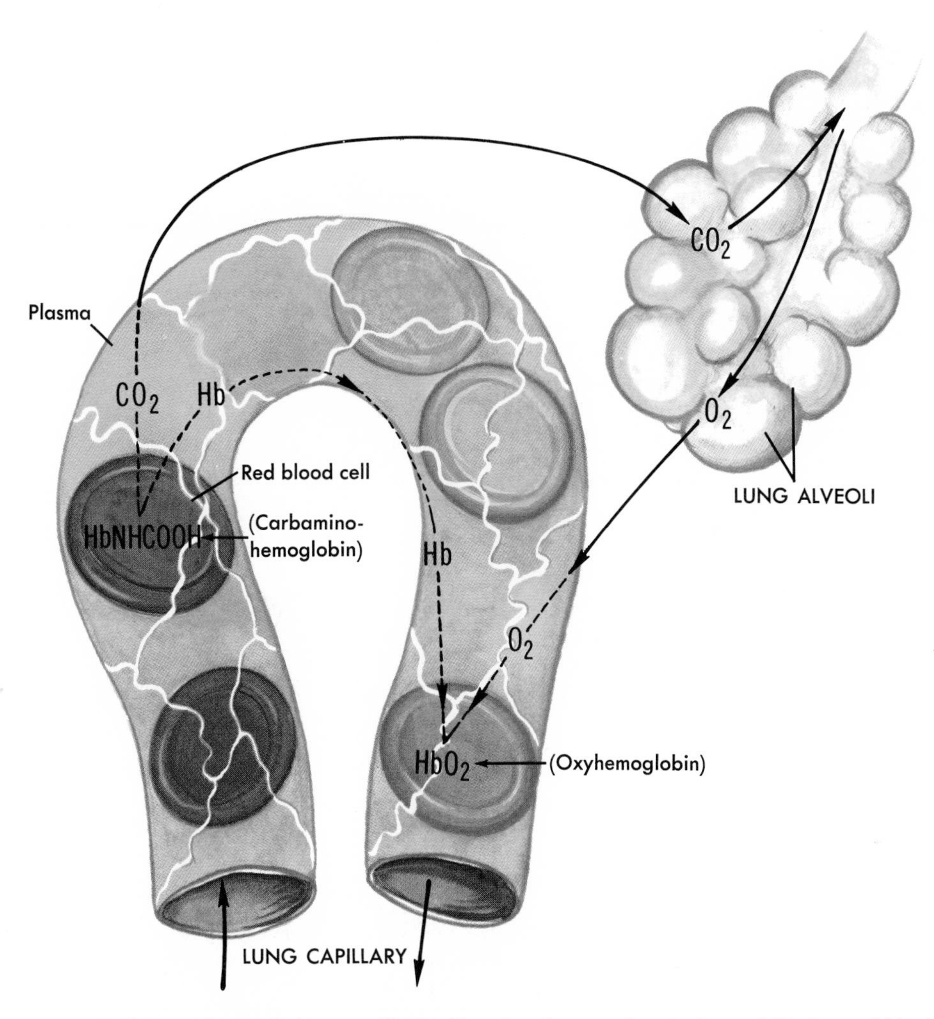

Fig. 7-2 Exchange of gases in lung capillaries. Drawing shows carbaminohemoglobin in a red blood cell dissociating to release CO_2, which diffuses out of blood into alveolar air. Simultaneously, O_2 is diffusing out of alveolar air into blood and associating with hemoglobin to form oxyhemoglobin. (From Anthony, C. P., and Kolthoff, N. J.: Textbook of anatomy and physiology, ed. 8, St. Louis, 1971, The C. V. Mosby Co.)

means increased breathing; more air moving in and out of the lungs per minute.) More carbon dioxide than normal would leave this patient's blood every minute. How would this affect the carbonic acid content of his blood? Increase it or decrease it? Do you think, therefore, that prolonged hyperventilation would tend to make his blood more or less alkaline than normal? In other words, would you expect this patient to develop alkalosis or acidosis?

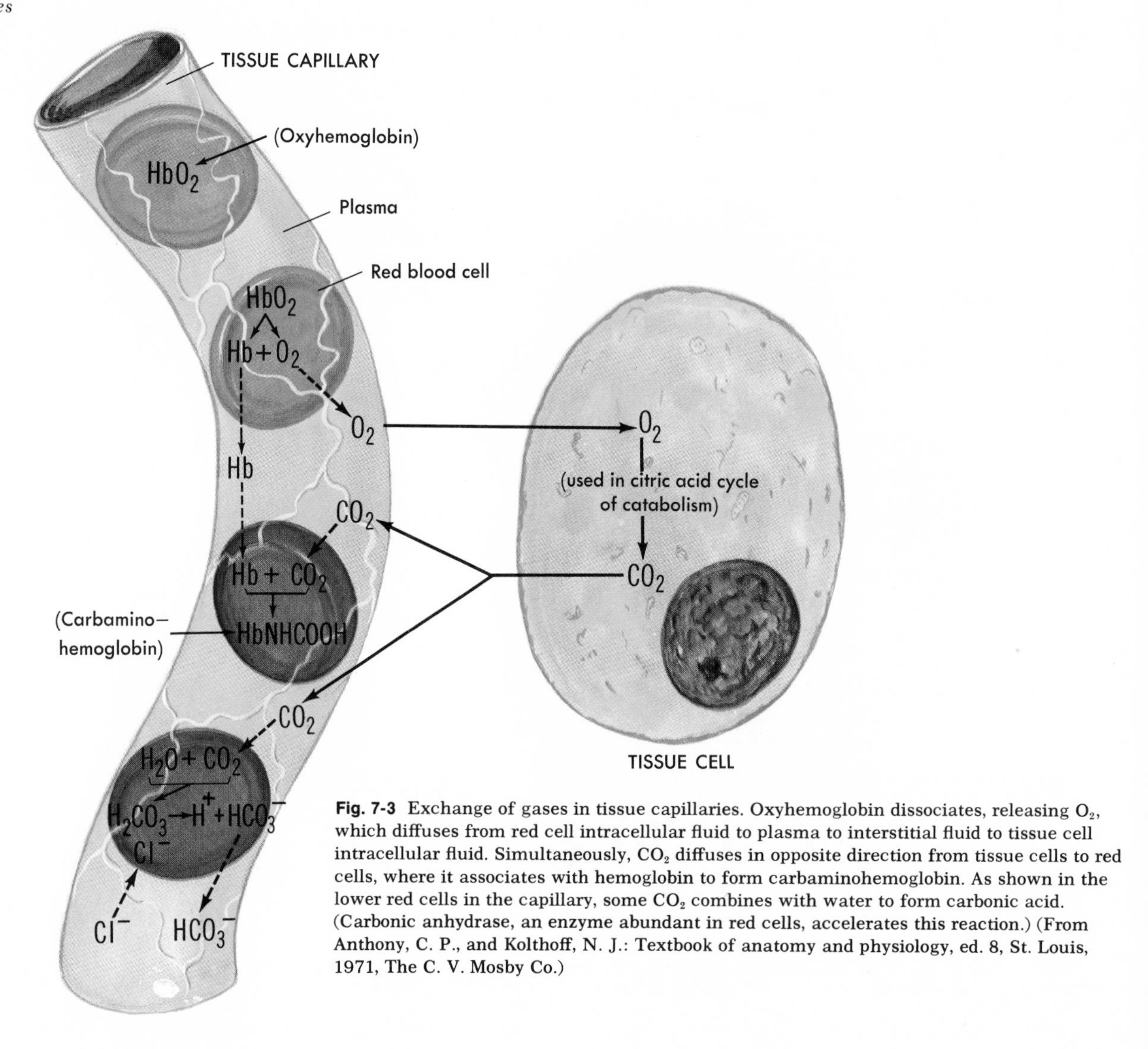

Fig. 7-3 Exchange of gases in tissue capillaries. Oxyhemoglobin dissociates, releasing O_2, which diffuses from red cell intracellular fluid to plasma to interstitial fluid to tissue cell intracellular fluid. Simultaneously, CO_2 diffuses in opposite direction from tissue cells to red cells, where it associates with hemoglobin to form carbaminohemoglobin. As shown in the lower red cells in the capillary, some CO_2 combines with water to form carbonic acid. (Carbonic anhydrase, an enzyme abundant in red cells, accelerates this reaction.) (From Anthony, C. P., and Kolthoff, N. J.: Textbook of anatomy and physiology, ed. 8, St. Louis, 1971, The C. V. Mosby Co.)

Exchange of gases in the tissues

The exchange of gases between the blood in tissue capillaries and the cells that make up the tissues is just the opposite of the exchange of gases between the blood in lung capillaries and the air in alveoli. As shown in Figure 7-3, in the tissue capillaries, oxyhemoglobin breaks down into oxygen and hemoglobin. Oxygen molecules move rapidly out of the blood through the tissue capillary membrane into the interstitial fluid and on into the cells that compose the tissues. While this is happening, carbon dioxide molecules are leaving the cells, entering the tissue capillaries, and uniting with hemoglobin molecules to form carbaminohemoglobin. In other words, the arterial blood that enters tissue capillaries becomes venous blood as it flows through them.

Mechanics of breathing

Breathing involves not only the organs of the respiratory system but also the brain, spinal cord, nerves, certain skeletal muscles, and even some bones and joints. In breathing, nerve impulses stimulate the diaphragm to contract, and as it contracts its shape changes. The domelike shape of the diaphragm flattens out so that it no longer protrudes into the chest cavity. The flattening of the diaphragm makes the chest cavity longer from top to bottom. Other muscle contractions raise the rib cage to make the chest cavity wider and greater in depth from front to back. As the chest cavity enlarges, the lungs expand along with it and air rushes into them and down into the alveoli. This part of respiration is called inspiration. For expiration to take place, the diaphragm and other respiratory muscles relax, and the chest cavity becomes smaller, thereby squeezing air out of the lungs.

Ordinarily we take about a pint (500 milliliters) of air into our lungs. Because this amount comes and goes regularly like the tides of the sea, it is referred to as the *tidal air*. The largest amount of air that we can breathe in and out in one inspiration and expiration is known as the *vital capacity;* in most adults this is about 4½ quarts (4,500 milliliters). A special device, the spirometer, is used to measure the amount of air exchanged in breathing. Physicians use this information mainly with patients who have lung or heart disease, because these conditions often lead to abnormal volumes of air being moved in and out of the lungs.

By now you know that the body uses oxygen to get energy for the work it has to do. Briefly, energy is stored in foods and the oxidation reactions of catabolism make this energy available for all kinds of cellular work (pp. 8 and 9). Therefore, the more work the body does, the more oxygen must be delivered to its millions of cells. One way this is accomplished is by increasing the rate and depth of respirations. Although we may take only 16 breaths a minute when we are not moving about, when we are exercising we take considerably more than this. Not only do we take more breaths; we also breathe in more air with each breath. Instead of about a pint of air, we may breathe deeply enough to take in several pints—sometimes even up to the limit of our vital capacity.

The way respirations are made to increase during exercise is an interesting example of the body's automatic regulation of its vital functions. When we are exercising, cells carry on catabolism at a faster rate than usual. This means that they form more carbon dioxide and that more carbon dioxide enters the blood in the tissue capillaries; therefore, the blood's carbon dioxide concentration starts to increase. This higher carbon dioxide concentration has a stimulating effect on the neurons of the respiratory center in the medulla. They send out more impulses to the respiratory muscles, and as a result, respirations increase, becoming both faster and deeper. Therefore, as more air moves in and out of the lungs per minute, more carbon dioxide leaves the blood, more oxygen enters it, and more energy becomes available for doing the extra work of exercise.

Another important automatic adjustment that helps supply cells with more oxygen when they are doing more work has to do with the circulatory system instead of the respiratory system. The heart pumps more blood through the body per minute because it beats faster and harder. This means that the millions of red blood cells make more round trips between the lungs and tissues each minute and so deliver more oxygen per minute to tissue cells.

Disorders of the respiratory system

Many different kinds of diseases and injuries may be responsible for respiratory disorders. For example, tuberculosis may destroy part of the lungs, lung cancer may do the same thing, and pneumonia may plug up alveoli. A brain hemorrhage may depress the respiratory center, causing respirations to become slow and labored or even to stop completely. In recent years, a disease rarely heard of in our grandfathers' generation—*emphysema*—has become increasingly common. In this disease the walls of many of the alveoli become greatly overstretched. They do not collapse during expiration as normal alveoli do, so air remains trapped in them instead of being exhaled. As a result, less air than normal is exhaled and inhaled, and less oxygen and carbon dioxide are exchanged between alveolar air and blood. Emphysema victims, therefore, develop *hypoxia.* A synonym for hypoxia is oxygen deficiency. If a person is hypoxic their cells receive less oxygen than what they need for normal functioning. Lung disease is not the only abnormality that can produce hypoxia. Anemia, for example, can also cause hypoxia. An anemic individual's red blood cells contain less hemoglobin than normal and so they transport less oxygen than normal.

outline summary

Basic plan similar to an inverted tree if it were hollow; leaves of tree would be comparable to alveoli, the microscopic sacs enclosed by networks of capillaries

NOSE

1 Structure
  a Nasal septum separates interior of nose into two cavities
  b Mucous membrane lines nose
  c Frontal, maxillary, sphenoid, and ethmoid sinuses drain into nose
2 Functions
  a Warms and moistens air inhaled
  b Contains sense organs of smell

PHARYNX

1 Structure
  a Two nasal cavities, mouth, esophagus, larynx, and eustachian tubes all have openings into pharynx
  b Tonsils and adenoids located in pharynx
  c Mucous membrane lines pharynx
2 Functions
  a Passageway for food and liquids
  b Passageway for air

LARYNX (voice box)

1 Structure
  a Several pieces of cartilage form framework
  b Mucous lining
  c Vocal cords stretch across interior of larynx
2 Functions
  a Passageway for air to and from lungs
  b Voice production

TRACHEA (windpipe)

1 Structure
  a Mucous lining
  b C-shaped rings of cartilage hold trachea open
2 Function—passageway for air to and from lungs

BRONCHI, BRONCHIOLES, AND ALVEOLI

1 Structure
  a Trachea branches into right and left bronchi
  b Each bronchus branches into smaller and smaller tubes called bronchioles
  c Bronchioles end in clusters of microscopic alveolar sacs, the walls of which are made up of alveoli
2 Function
  a Bronchi and bronchioles—passageway for air to and from alveoli
  b Alveoli—exchange of gases between air and blood

LUNGS

1 Structure
  a Size—large enough to fill chest cavity except for middlespace occupied by heart and large blood vessels
  b Apex—narrow upper part of each lung, under collarbone
  c Base—broad lower part of each lung; rests on diaphragm
  d Pleura—moist, smooth, slippery membrane that lines chest cavity and covers outer surface of lungs; prevents friction between lungs and chest wall during breathing
2 Function—breathing (respiration)

RESPIRATION

Exchange of gases in lungs

1 Carbaminohemoglobin breaks down into carbon dioxide and hemoglobin
2 Carbon dioxide moves out of lung capillary blood into alveolar air and out of body in expired air
3 Hemoglobin combines with oxygen, producing oxyhemoglobin

Exchange of gases in tissues

1 Oxyhemoglobin breaks down into oxygen and hemoglobin
2 Oxygen moves out of tissue capillary blood into tissue cells
3 Carbon dioxide moves from tissue cells into tissue capillary blood
4 Hemoglobin combines with carbon dioxide, forming carbaminohemoglobin

Mechanics of breathing

1 Contraction of diaphragm and of chest-elevating muscles enlarges chest cavity, expands lungs, and causes air to move down into lungs
2 Relaxation of diaphragm and of chest elevators decreases size of chest cavity, deflates lungs, and causes air to move out of lungs
3 Tidal air—amount normally breathed in and out with each breath
4 Vital capacity—largest amount of air we can breathe in and out with one breath
5 Rate—usually about 16 to 20 breaths a minute; much faster during exercise

DISORDERS OF THE RESPIRATORY SYSTEM

1 Tuberculosis
2 Lung cancer
3 Pneumonia
4 Brain hemorrhage
5 Emphysema

new words

carbaminohemoglobin
emphysema
hyperventilation
hypoxia
oxyhemoglobin
pleurisy
pneumothorax
respiration
tidal air
thoracentesis
vital capacity

review questions

1 What and where are the pharynx and larynx?
2 What and where are alveoli and the pleura?
3 Why does sinusitis frequently follow a common cold?
4 Where are the tonsils and adenoids located?
5 What and where is the "Adam's apple"?
6 What tubes connect each middle ear with the pharynx?
7 What does the term "vital capacity" mean?

8

The circulatory system

Have you ever thought what would happen if all the transportation ceased in your city or town? How soon you would have no food to eat and how soon rubbish and waste would pile up, for instance? Stretch your imagination just a little and you can visualize many disastrous results. Lack of food transportation alone would threaten every individual's life. Similarly, cells, the individuals of the body, face a threat to survival if transportation within the body ceases. The system that provides this vital transportation service is the circulatory system. Its organs are the heart, blood vessels, lymphatic vessels, and lymph nodes. The fluids circulated are blood and lymph. We shall begin this chapter with a discussion of blood.

Blood structure and functions

Blood is one of the three main fluids of the body. (The fluid around cells and the fluid inside them are the other two.) The liquid part of blood is called *plasma*. Floating in the plasma are hundreds of millions of blood cells.

There are three main types and several subtypes of blood cells as follows:

1. Red cells, or erythrocytes
2. White cells, or leukocytes
   a. Granular leukocytes (have granules in their cytoplasm)
      (1) Neutrophils
      (2) Eosinophils
      (3) Basophils
   b. Nongranular leukocytes (do not have granules in their cytoplasm)
      (1) Lymphocytes
      (2) Monocytes

Examine Figure 8-1 to see what each of these different kinds of blood cells looks like under the microscope.

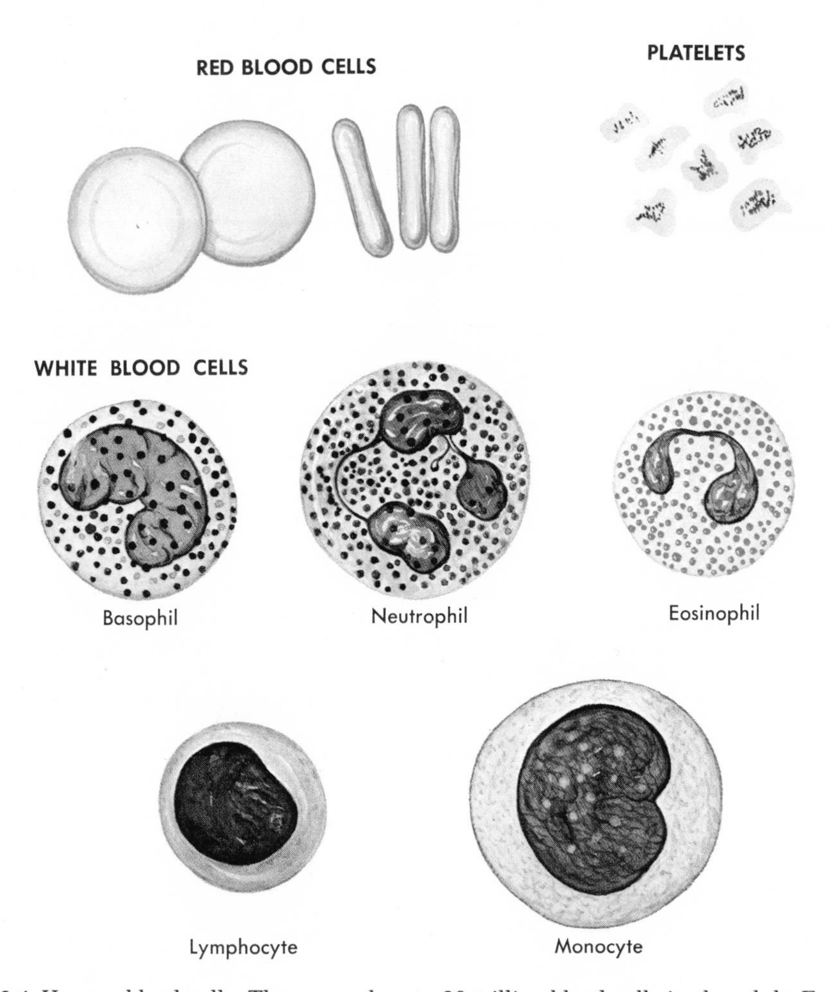

Fig. 8-1 Human blood cells. There are close to 30 trillion blood cells in the adult. Each cubic millimeter of blood contains from 4½ to 5½ million red blood cells and an average total of 7,500 white blood cells.

The number of blood cells in the body is hard to believe. For instance, 5,000,000 red cells and 7,500 white blood cells in 1 cubic millimeter of blood (a drop only about 1/25 of an inch long and wide and high) would be considered a normal "red count" and a normal "white count." Since both red and white cells are continually being destroyed, the body has to continually make new ones to take their place at a really staggering rate—a few million red cells alone each second.

Two kinds of connective tissue—*myeloid tissue* and *lymphatic tissue*—make blood cells for the body. A better-known name for myeloid tissue is red bone marrow. In the adult body it is present chiefly in the sternum, ribs, and cranial bones, although a few of the other bones also contain small amounts of this valuable substance. Red bone marrow forms all types of blood cells except lymphocytes and monocytes. These are formed by lymphatic tissue located chiefly in the lymph nodes and the spleen.

Sometimes for one reason or another, the blood-forming tissues cannot maintain normal numbers of blood cells. The number of red cells, for example, frequently drops below normal, a process resulting in a condition that is called *anemia. Leukopenia,* or an abnormally low white count, occurs only occasionally. An abnormally high white count, *leukocytosis,* however, is quite common, since it almost always accompanies infections. There is also a malignant disease, *leukemia,* in which the number of white cells increases tremendously; you may have heard of this disease as "blood cancer."

Blood cells perform several important functions. Red cell functions were discussed in the preceding chapter. They transport oxygen to all the other cells in the body. The chemical in them called *hemoglobin* unites with oxygen to form oxyhemoglobin. If hemoglobin falls below the normal level, as it does in anemia, it starts an unhealthy chain reaction: less hemoglobin – less oxygen transported to cells – slower catabolism by cells – less energy supplied to cells – decreased cellular functions. If you understand this relation between hemoglobin and energy, you can guess correctly what an anemic person's chief complaint will probably be – that he feels "so tired all the time." Red blood cells perform one other function; besides transporting oxygen, they also transport carbon dioxide.

White blood cells carry on a function perhaps slightly less vital than red cells, but one, nevertheless, that often has lifesaving importance. They defend the body from microorganisms that have succeeded in invading the tissues or bloodstream. Neutrophils and monocytes, for example, engulf microbes. They actually take them into their own cell bodies and digest them. This process is called *phagocytosis,* and the cells that digest microbes are called *phagocytes.* Most numerous of the phagocytes are the neutrophils, but in addition to neutrophils and monocytes, the body also has other phagocytes that are not white blood cells. They are classed as reticuloendothelial cells – a type of connective tissue cells. Microglia (p. 53), for example, are reticuloendothelial phagocytes.

Lymphocytes, it is now thought, also help protect us against infections. But they do it by a process different from phagocytosis. Lymphocytes are believed to function in the immune mechanism, the process that makes us immune to infectious diseases. It is thought that the immune mechanism starts to operate when microbes invade the body. In some way or another their presence acts to stimulate lymphocytes to start multiplying and to become transformed into plasma cells. Presumably, each kind of microbe can stimulate only one specific kind of lymphocyte to multiply and form one kind of plasma cell. The kind of plasma cell formed is the kind that can make a specific kind of antibody. The kind of antibody made is the one that can destroy the particular kind of microbe that has invaded the body in the first place and set the immune mechanism in operation.

Blood plasma

Blood plasma is the liquid part of the blood, or blood minus its cells. It consists of water with many substances dissolved in it. All of the chemicals needed by cells to stay alive have to be brought to them by the blood. Food and salts, therefore, are dissolved in plasma. So, too, is a small amount of oxygen. (Most of the oxygen in the blood is carried in the red blood cells as oxyhemoglobin.) Wastes that cells must get rid of are dissolved in plasma and transported to the excretory organs. And, finally, dissolved in plasma are the hormones that help control our cells' activities and the antibodies that help protect us against microorganisms.

Many people seem curious about how much blood they have. The amount depends upon how big they are and whether they are male or female. A big person has more blood than a small person and a man has more blood than a woman. But as a general rule, most adults probably have between 4 and 6 quarts (4,000 and 6,000 milliliters) of blood.

The volume of the plasma part of blood is usually a little more than half the volume of whole blood. Blood cells make up the remaining part of whole blood's volume. Examples of normal volumes are the following: plasma volume–2,600 milliliters; blood cell volume–2,400 milliliters; total blood volume–5,000 milliliters.

If you read many advertisements or watch many television commercials, you may think that almost everyone has "acid blood" at some time or other. Nothing could be farther from the truth. Blood is alkaline; it rarely reaches even the neutral point. If the alkalinity of your blood decreases toward neutral, you are a very sick person; in fact, you have what is called *acidosis*. But even in this condition, blood almost never becomes the least bit acid; it just becomes less alkaline than normal.

Blood types (or blood groups)

Before we discuss blood types, we need to define the terms "antigens" and "antibodies." An *antigen* is a substance that can stimulate the body to make antibodies. Almost all substances that act as antigens are foreign proteins. In other words, they are not the body's own natural proteins, but are proteins that have entered the body from the outside–by injection or transfusion or some other method.

The word "antibody" can be defined in terms of what causes its formation or in terms of how it functions. Defined the first way, an *antibody* is a substance made by the body in response to stimulation by an antigen. Defined according to its functions, an antibody is a substance that reacts in some way with the antigen that stimulated its formation. Many antibodies, for example, clump (or agglutinate) their antigens; that is, they cause their antigens to stick together in little clusters.

Every person's blood belongs to one of the following four blood types: type A, type B, type AB, or type O. Suppose that you have type A blood (as do about 41% of Americans). The letter A stands for a certain type of antigen (a protein) present in your red blood cells when you were born. Because you were born with type A antigen, your body does not form antibodies to react with it. In other words, your blood plasma contains no anti-A antibodies. It does contain anti-B antibodies, however. For some unknown reason, these antibodies are present naturally in type A blood. The body did not form them in response to the presence of their antigen. In summary, then, in type A blood, the red cells contain type A antigen and the plasma contains anti-B antibodies.

In type AB blood, as its name indicates, the red cells contain both type A and type B antigens and the plasma contains neither anti-A nor anti-B antibodies. The opposite is true of type O blood–its red cells contain neither type A nor type B antigens and its plasma contains both anti-A and anti-B antibodies.

Bad effects or even death can result from a blood transfusion if the donor's red cells become agglutinated by antibodies in the recipient's plasma. If a donor's red cells do not contain any A or B antigen, they of course cannot be clumped by anti-A or anti-B antibodies. For this reason the type of blood that contains neither A nor B antigens—namely, type O blood—can be used as donor blood without the danger of anti-A or anti-B antibodies clumping its red cells. Type O blood is therefore called *universal donor* blood. *Universal recipient* blood is type AB; it contains no anti-A or anti-B antibodies in its plasma, and so cannot clump any donor's red cells containing A or B antigens.

In recent years, the expression *Rh positive* blood has become a familiar one. It means that the red blood cells of this type blood contain an antigen called the Rh factor. If, for example, a person has type AB, Rh positive blood, his red cells contain type A antigen, type B antigen, and the Rh factor.

In *Rh negative* blood, the red cells do not contain the Rh factor and the plasma does not normally contain anti-Rh antibodies. But if Rh positive blood cells are introduced into an Rh negative person's body, anti-Rh antibodies soon appear in his blood plasma. In this fact lies the danger for a baby born to an Rh negative mother and Rh positive father. If the baby is Rh positive, the mother's body may be stimulated to form anti-Rh antibodies. If she later carries another Rh positive fetus, it may develop a disease called erythroblastosis fetalis, caused by the mother's Rh-antibodies reacting with the baby's Rh positive cells.

Blood clotting

Your life might someday be saved just because your blood can clot. A clot plugs up torn or cut vessels and so stops bleeding that otherwise might prove fatal. The story of how blood clots is the story of a rapid-fire reaction. The first step in the chain is some kind of an injury to a blood vessel that makes a rough spot in its lining. (Normally the lining of blood vessels is extremely smooth.) Almost immediately some of the blood platelets break up as they flow over the rough spot in the vessel's lining and release a substance into the blood, which leads to the formation of other substances called *platelet factors* (Figure 8-2). In the next step platelet factors combine with prothrombin (a protein present in normal blood), calcium, and other substances to form *thrombin*. Then in the last step, thrombin reacts with fibrinogen (another protein present in normal blood) to change it to a gel called fibrin. Under the microscope fibrin looks like a tangle of fine threads with red blood cells caught in the tangle. The red cells give the red color to clotted blood.

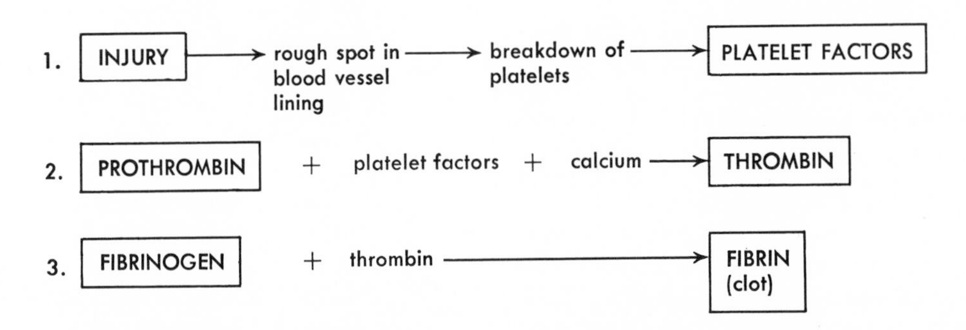

Fig. 8-2 Diagram showing the main steps in blood clotting—a process far more complex than is indicated here.

The clotting mechanism as just described contains clues for ways to stop bleeding by speeding up blood clotting. For example, you might simply apply gauze to a bleeding surface. Its slight roughness would cause more platelets to break down and release more platelet factors. The additional platelet factors would then make the blood clot more quickly.

Physicians frequently prescribe vitamin K before surgery to make sure that the patient's blood will clot fast enough to prevent hemorrhage. Figure 8-3 shows the somewhat roundabout way in which vitamin K acts to hasten blood clotting.

Unfortunately, clots sometimes form in uncut blood vessels of the heart, brain, lungs, or some other organ—a dreaded thing, because clots may produce sudden death by shutting off the blood supply to a vital organ. When a clot stays in the place where it formed, it is called a *thrombus*, and the condition is spoken of as *thrombosis*. If part of the clot dislodges and circulates through the bloodstream, the dislodged part is then called an *embolus* and the condition is called an *embolism*. Suppose that your doctor told you that you had a clot in one of your coronary arteries. Which diagnosis would he make—coronary thrombosis or coronary embolism—if he thought that the clot had formed originally in one of the small vessels that supply blood to heart muscle cells? Doctors now have some drugs that they can use to help prevent thrombosis and embolism. Dicumarol is one. It blocks the stimulating effect of vitamin K on the liver, and consequently the liver cells make less prothrombin. The blood prothrombin content soon falls low enough to prevent clotting (Figure 8-2).

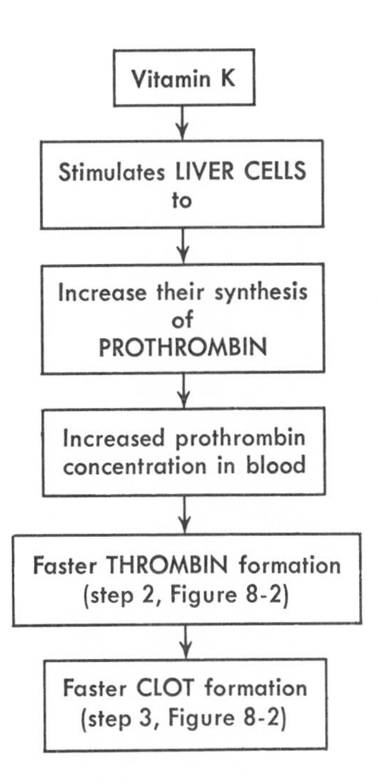

Fig. 8-3 Diagram showing how vitamin K acts to accelerate blood clotting.

Heart

No one needs to be told where his heart is or what it does. Everyone knows that the heart is in the chest, that it beats night and day to keep the blood flowing, and that if it stops, life stops. Most of us probably think of the heart as located on the left side, but as you can see in Figure 1-2, it occupies the lower portion of the mediastinum with most of its mass to the left of the midline of the body. The *apex* (blunt point of the lower edge of the heart) lies on the diaphragm, pointing toward the left. To count the apical beat, one must place a stethoscope directly over the apex, that is, in the space between the fifth and sixth ribs on a line with the midpoint of the left clavicle.

If you cut open a heart you can see many of its main structural features (Figure 8-4). It is hollow, not solid. A partition (the

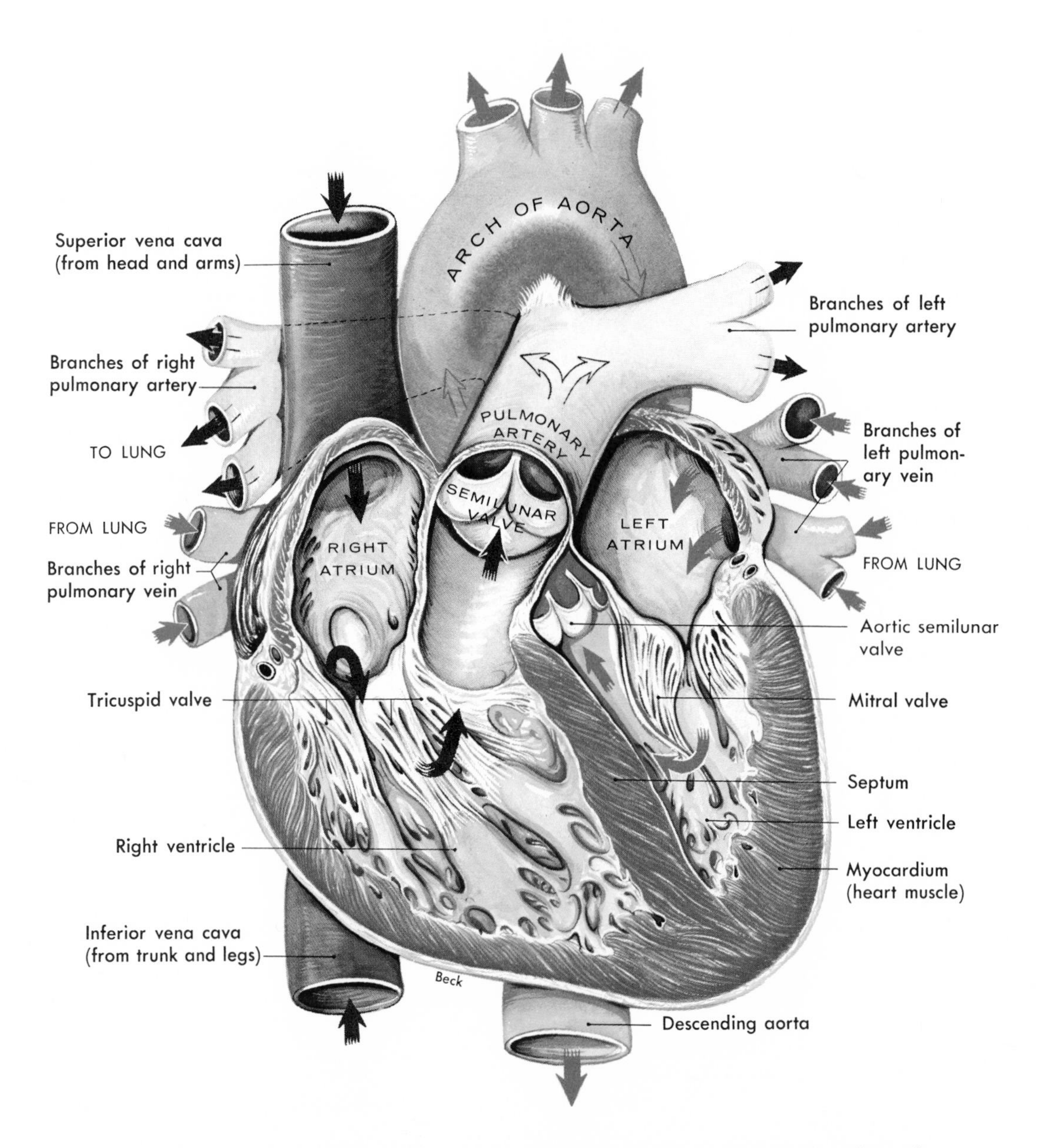

Fig. 8-4 Cutaway view of the front section of the heart showing the four chambers, the valves, the openings, and the major vessels. Arrows indicate the direction of blood flow; the black arrows represent unoxygenated blood and the red arrows oxygenated blood. The two branches of the right pulmonary vein extend from the right lung behind the heart to enter the left atrium.

septum) divides it into right and left sides. It has four cavities inside: a small upper cavity *(atrium)* and a larger lower cavity *(ventricle)* on each side. Cardiac muscle tissue composes the wall of the heart; it is usually referred to as the *myocardium.* Myocarditis, therefore, means inflammation of the heart muscle. A very smooth tissue, *endocardium,* lines the cavity of the heart. Endocarditis, of course, means inflammation of the heart lining. This condition can cause rough spots to develop in the endocardium, which may lead to thrombosis. (Look back at Figure 8-2 to find out why.)

The heart has a covering as well as a lining. Its covering, the *pericardium,* consists of two layers of fibrous tissue with a small space in between. The inner layer of the pericardium covers the heart like an apple skin covers an apple, but the outer layer fits around the heart like a loose-fitting sack, allowing enough room for the heart to beat in it. These two pericardial layers slip against each other without friction when the heart beats because they are moist, not dry surfaces. (Have you ever walked carefully on a wet floor? If so, you were consciously or unconsciously using your knowledge of the principle that moist surfaces are slippery.) A thin film of pericardial fluid furnishes the lubricating moistness between the heart and its enveloping pericardial sac.

Four valves keep blood flowing through the heart in the right direction. The *tricuspid valve,* at the opening between the right atrium and the right ventricle, lets blood flow from the atrium into the ventricle but prevents it from flowing in the opposite direction. The *mitral* or *bicuspid valve,* at the opening between the left atrium and the left ventricle, serves the same purpose for the left side of the heart. The *pulmonary semilunar valve* at the beginning of the pulmonary artery allows blood to flow out of the right ventricle but prevents it from backflowing into the ventricle. The *aortic semilunar valve* at the beginning of the aorta allows blood to flow out of the left ventricle up into the aorta but prevents backflow into this ventricle.

Embedded in the wall of the heart are four structures that conduct impulses through the heart muscle to cause first the atria and then the ventricles to contract. The names of the structures that make up this conduction system of the heart are the *sinoatrial node* (SA node, or the pacemaker of the heart), *atrioventricular node* (AV node), the *bundle of His,* (or AV bundle), and the *Purkinje fibers.* Impulse conduction normally starts in the heart's pacemaker, namely, the SA node. From here it spreads, as you can see in Figure 8-5, in all directions through both atria. This causes the atria to contract. When impulses reach the AV node, it relays them by way of the bundle of His and Purkinje fibers to the ventricles, causing them to contract. Normally, therefore, a ventricular beat follows each atrial beat; but various diseases can damage the heart's conduction system and thereby disturb the rhythmical beating of the heart. One such disturbance is the condition commonly called *heartblock.* Impulses are blocked from getting through to the ventricles with the result that the heart beats at a much slower rate than normal. A physician may treat heartblock by implanting in the heart an *artificial pacemaker,* an electrical device that causes ventricular contractions at a rate fast enough to maintain an adequate circulation of blood.

Blood vessels

Arteries, veins, and capillaries—these are the names of the three main kinds of blood vessels. *Arteries* carry blood away

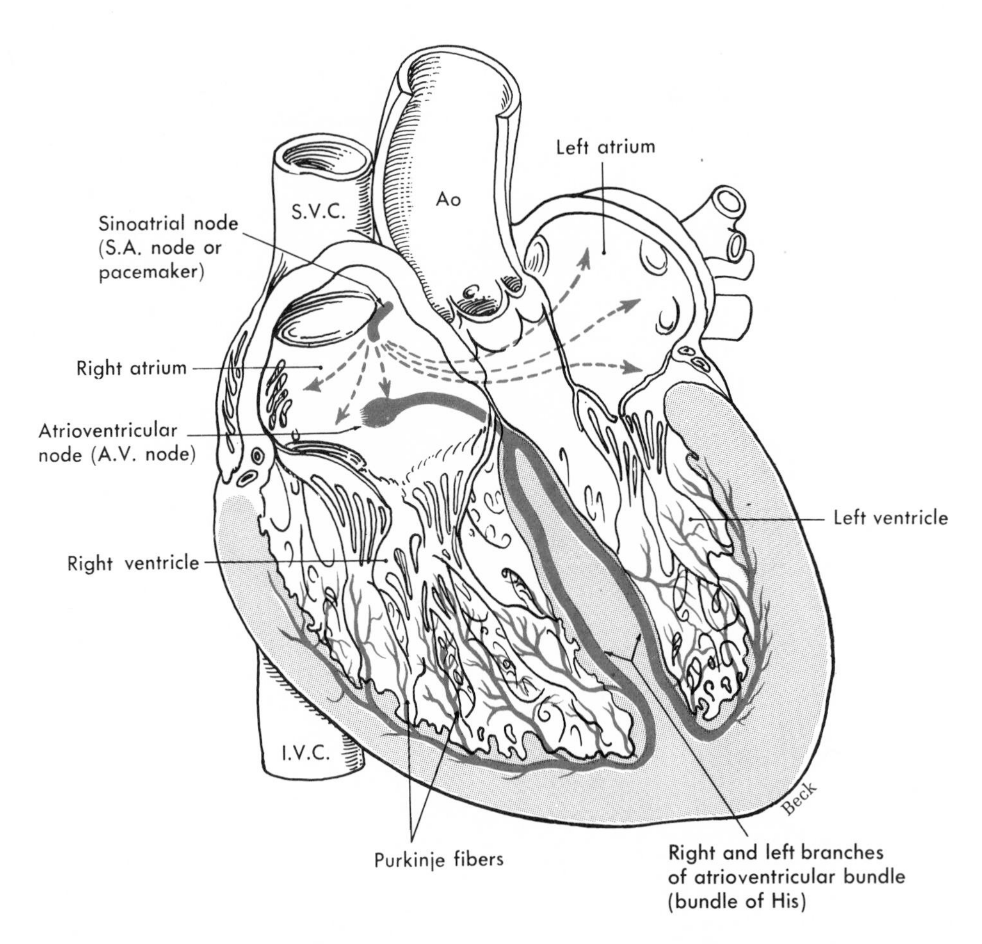

Fig. 8-5 The conduction system of the heart. The sinoatrial node in the wall of the right atrium sets the basic pace of the heart's rhythm so it is called the "pacemaker." (From Anthony, C. P., and Kolthoff, N. J.: Textbook of anatomy and physiology, ed. 8, St. Louis, 1971, The C. V. Mosby Co.)

from the heart toward capillaries. *Veins* carry blood toward the heart away from capillaries. *Capillaries* carry blood from tiny arteries *(arterioles)* into tiny veins *(venules)*. When a surgeon cuts into the body, he can see arteries, arterioles, veins, and venules. He cannot see capillaries, because they are microscopic-sized vessels. An important point about the structure of a capillary is that its wall (that is, the capillary membrane) is very thin – only one cell thick to be exact. Because a single layer of flat epitheliallike cells composes the capillary membrane, substances such as foods, oxygen, and wastes can quickly pass through it on their way to or from cells. Figures 8-6 and 8-7 illustrate the structure of arteries, capillaries, and veins.

The largest artery in the body is the *aorta*. The largest vein is the *vena cava*. The aorta carries blood out of the left ventricle of the heart, and the vena cava returns blood to the right atrium after the blood has circulated through the body.

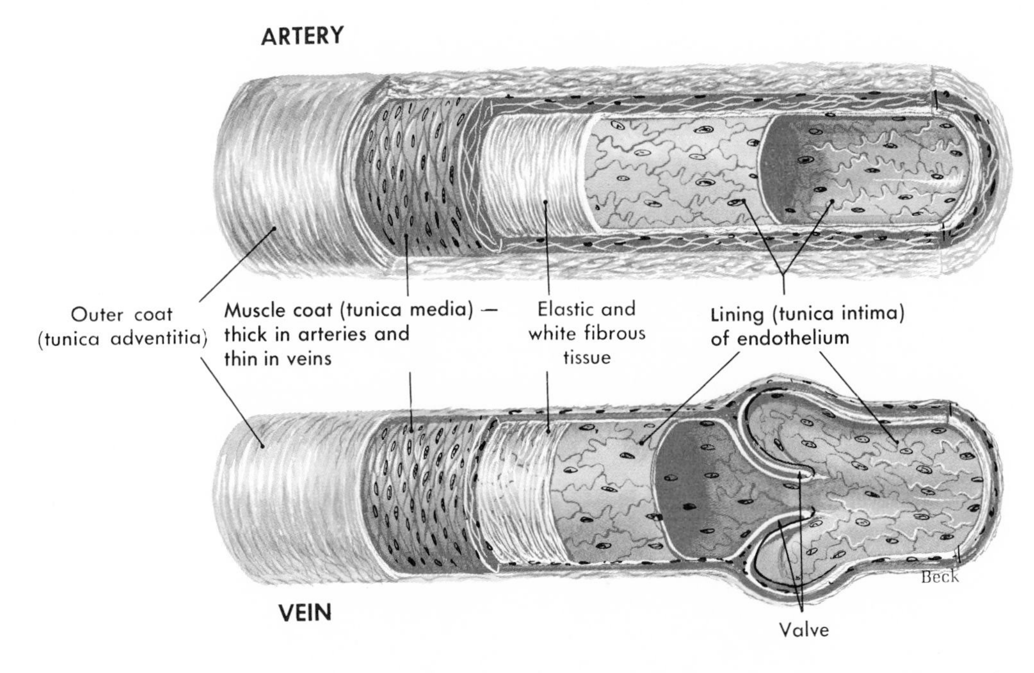

Fig. 8-6 Schematic drawings of an artery and vein showing comparative thicknesses of the three coats: the outer coat (tunica adventitia), the muscle coat (tunica media), and the lining of endothelium (tunica intima). Note that the muscle and outer coats are much thinner in veins than in arteries and that veins have valves.

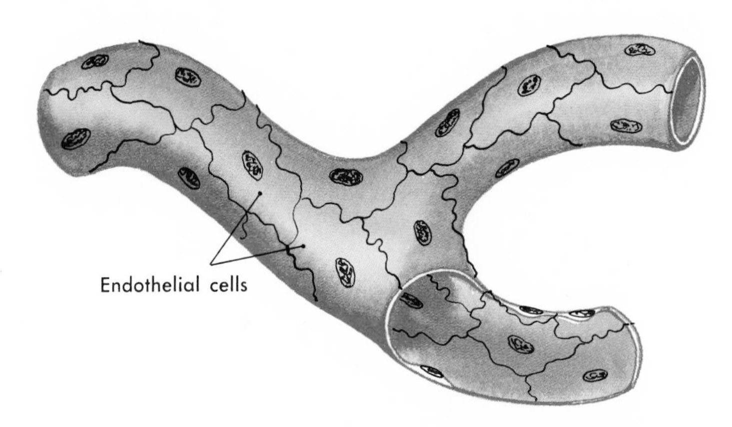

Fig. 8-7 The walls of capillaries consist of only a single layer of endothelial cells. These thin, flattened cells permit the rapid movement of substances between blood and interstitial fluid. Note that capillaries have no smooth muscle layer, elastic fibers, or surrounding coats.

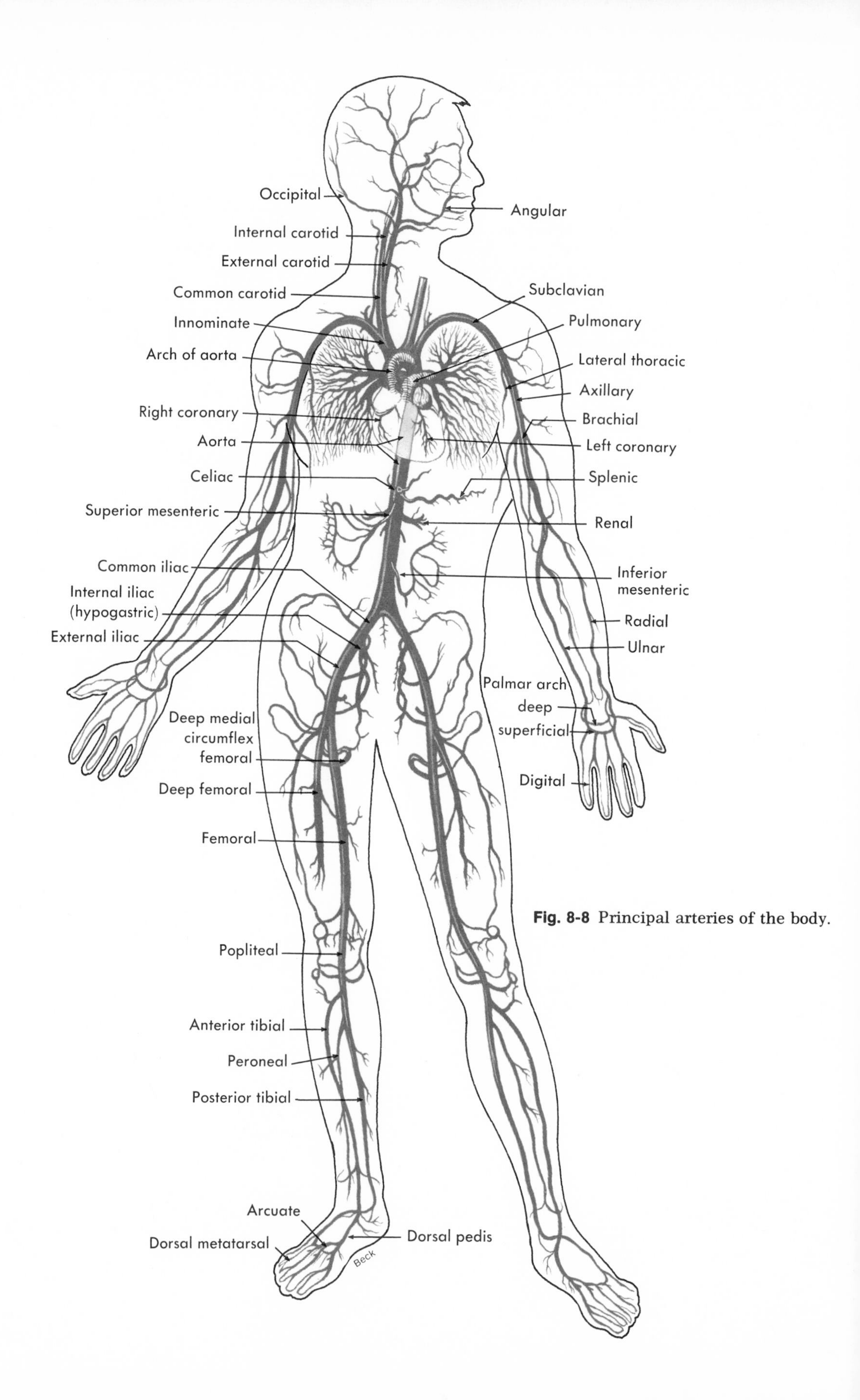

Fig. 8-8 Principal arteries of the body.

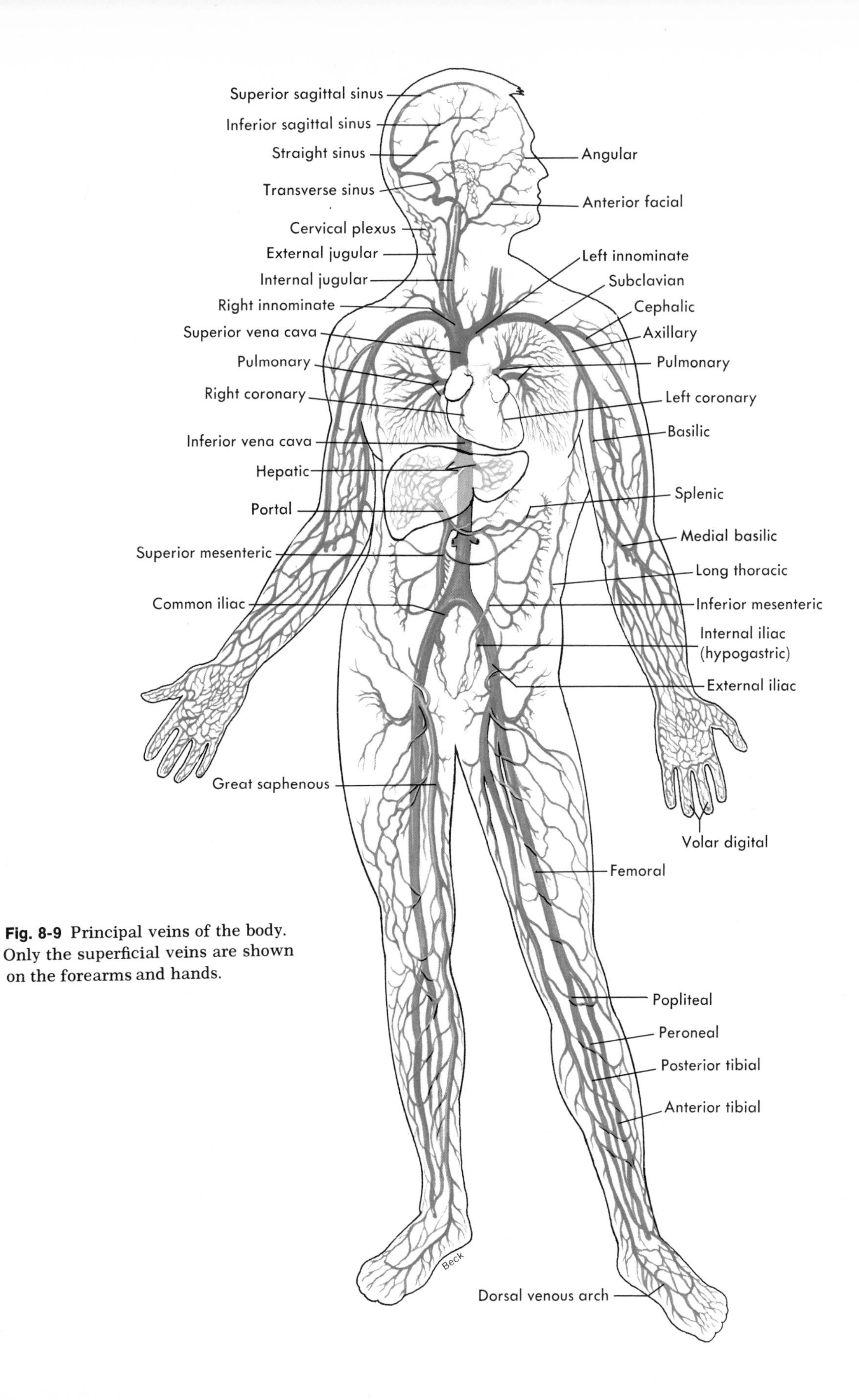

Fig. 8-9 Principal veins of the body. Only the superficial veins are shown on the forearms and hands.

Study Figure 8-8 to find out the names of the main arteries of the body and Figure 8-9 for the names of the main veins. Then see whether you can answer these questions: what is the name of the main artery of the thigh? of the upper arm? of the thumb side of the lower arm? What are neck arteries called? neck veins? Answer question 23, p. 136. Make up some more questions to quiz yourself about blood vessel names and locations.

Circulation

The word circulation implies that something moves, that it moves over a circular route, and that it moves over this route repeatedly. Applied to blood, circulation means the movement of blood over and over again through vessels that form a circular route. (By circular route we mean one that begins and ends at the same place, not necessarily one shaped like a circle.) Since the heart pumps the blood and since the blood returns to the right atrium of the heart when it has completed one circuit through the blood vessels, we shall consider that circulation starts in the right atrium. Examine Figure 8-11 carefully. Imagine that you have injected a drug into a patient's right upper arm. The drug would soon diffuse into the blood in the capillaries of the upper arm. Look again at Figure 8-11 to find out what kind of vessels the blood would flow into as it left the arm's capillaries. Notice next what two organs the drug-containing blood would have to pass through before it could enter the capillaries of any other organ in the body. First it would have to return to the heart (which side? which chamber?). Next it would have to circulate through the lungs, come back to the heart (which side? which chamber?), and then be pumped by the left ventricle out into the aorta and

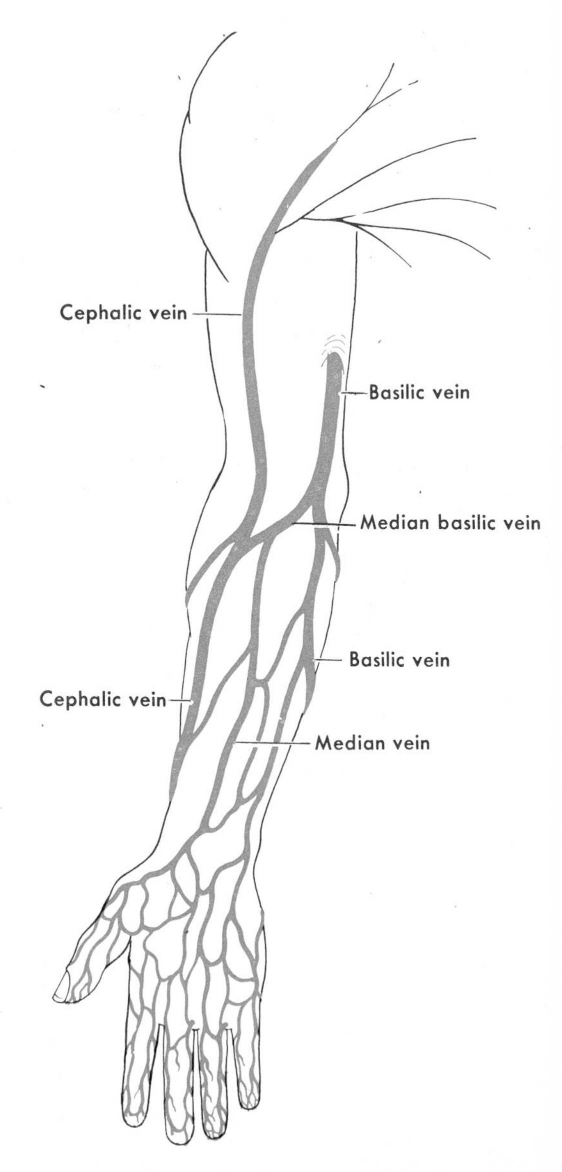

Fig. 8-10 Main superficial veins of the arm.

on into arteries, arterioles, and capillaries of other organs. Make sure that you understand the blood's circulation route by answering questions 14 to 18 on p. 136 and checking your answers with Figure 8-11.

As blood flows through capillaries in the lungs, it changes from venous blood to arterial blood by unloading carbon dioxide

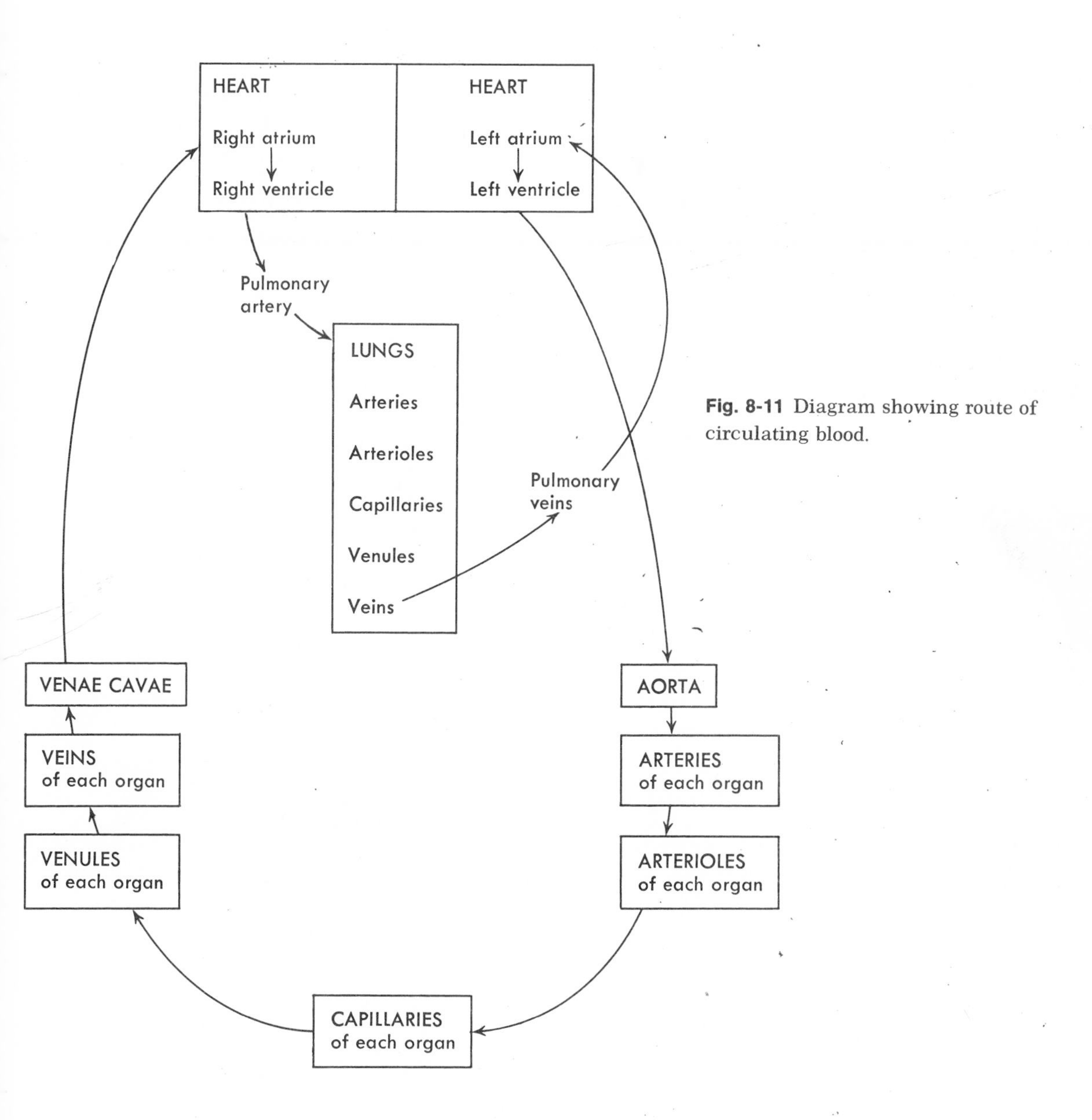

Fig. 8-11 Diagram showing route of circulating blood.

and picking up oxygen. Its color changes in the process from the deep crimson that identifies venous blood to the bright scarlet of arterial blood. Then as blood flows through tissue capillaries (all capillaries except those in the lungs), it changes back from arterial to venous blood by oxygen leaving the blood to enter cells and carbon dioxide leaving the cells to enter blood.

Blood flows into the heart muscle itself by way of two small vessels that are surely the most famous of all the blood vessels—the coronary arteries—famous because coronary disease kills so many thousands every year. The coronary arteries are the aorta's first branches. The openings into these small vessels lie behind the flaps of the aortic semilunar valves. In both coronary thrombosis and coronary embolism (p. 119), a blood clot occludes or plugs up some part of a coronary artery. Blood cannot pass through the occluded vessel and so cannot reach the heart muscle cells it normally supplies. Deprived of oxygen, these cells soon die; or to use the medical term, myocardial infarction occurs. Myocardial infarction is a common cause of death in middle-aged and old people.

The term "portal circulation" refers to the following route of blood flow through the liver. Veins from the abdominal digestive organs (stomach, pancreas, and intestines) and from the spleen empty into the portal vein. Branches of the portal vein empty into arterioles, which empty into the capillaries of the liver. Blood leaves the liver by way of the hepatic vein, which drains into the inferior vena cava. How does portal circulation differ from circulation through other organs? Does venous blood ordinarily flow into capillaries? See Figure 8-11 if you are not sure about this. The main reason for the detour of venous blood through the liver before its return to the heart is probably the process of liver glycogenesis (p. 95).

Circulation in a baby before birth is not the same as after birth, mainly because before birth the baby's blood has to get oxygen and food from the mother's blood instead of from its own lungs and digestive tract. Therefore, relatively little blood circulates through the baby's lungs and digestive organs before it is born, but all of the baby's blood circulates through the placenta (afterbirth) where the exchange of substances with the mother's blood occurs. As you read the next paragraph, refer often to Figure 8-12.

Two *umbilical arteries* (extensions of the baby's internal iliac arteries) carry blood to the *placenta*, a highly vascular structure attached to the inside of the mother's uterus. As the baby's blood circulates through the placenta, oxygen and foods enter it from the mother's blood while wastes move from the baby's to the mother's blood. After flowing through the placenta, blood returns to the baby's body by way of one *umbilical vein*. (Note in Figure 8-12 the red color of this vessel to indicate oxygenated blood.) Part of the returning blood circulates through the liver; the rest flows through the *ductus venosus* into the inferior vena cava, and on into the right atrium of the baby's heart. Some of the blood then flows to the lungs

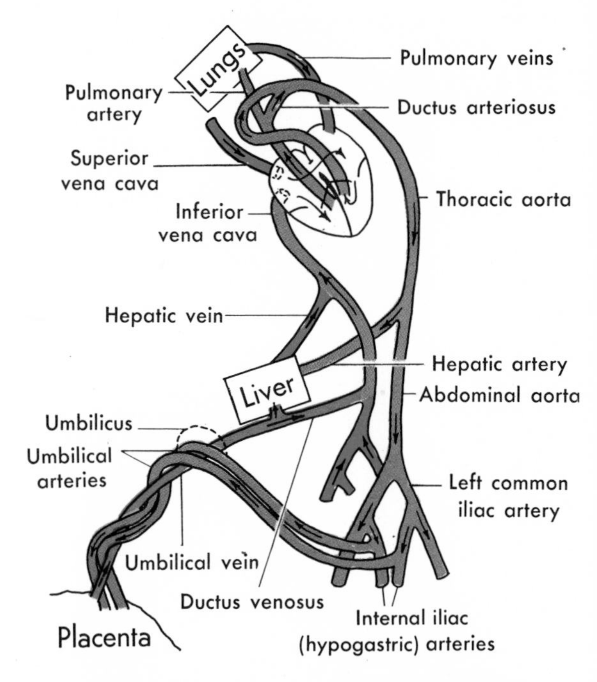

Fig. 8-12 Diagram of fetal circulation. Note these six structures that are present in a baby's body at birth, but not in a normal adult body: (1) umbilical arteries (two of them); (2) placenta (afterbirth); (3) umbilical vein; (4) ductus venosus; (5) ductus arteriosus; (6) foramen ovale (the opening—not labeled—between the right and left atria of the heart).

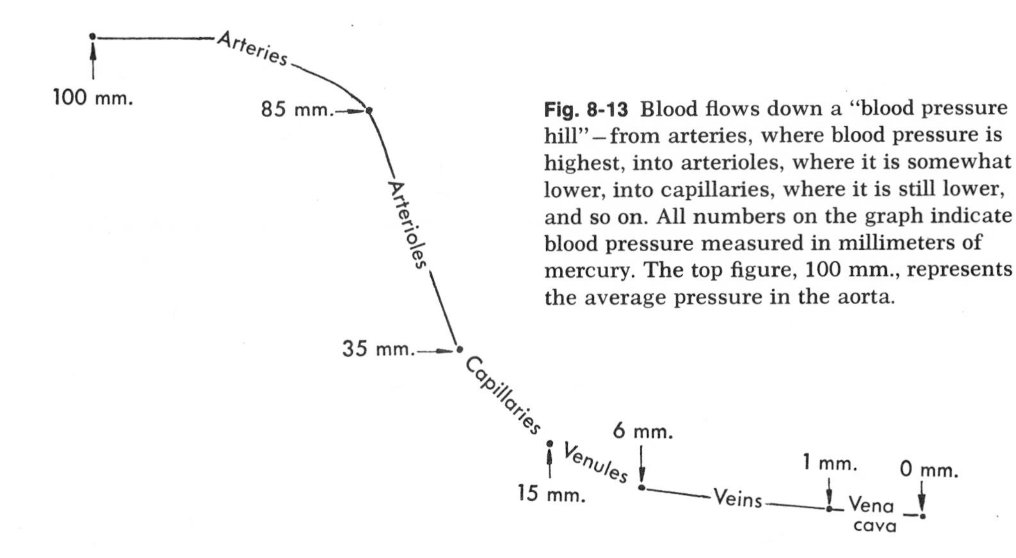

Fig. 8-13 Blood flows down a "blood pressure hill"—from arteries, where blood pressure is highest, into arterioles, where it is somewhat lower, into capillaries, where it is still lower, and so on. All numbers on the graph indicate blood pressure measured in millimeters of mercury. The top figure, 100 mm., represents the average pressure in the aorta.

as it does in the adult body. But some of it flows through two detours that bypass the lungs. One is a hole *(foramen ovale)* in the septum of the heart between the two atria. The other is a small vessel (the *ductus arteriosus*) that leads from the pulmonary artery to the thoracic aorta. After a baby is delivered and the umbilical cord is cut, the umbilical arteries, umbilical vein, and placenta no longer function. Soon the foramen ovale closes, and eventually the ductus venosus and arteriosus become fibrous cords.

Blood pressure

Perhaps a good way for us to try to understand blood pressure is to try to answer the questions what? where? why? and how? about it. What is blood pressure? Just what the words say—blood pressure is the pressure, or push, of blood.

Blood pressure exists in all blood vessels, but it is highest in the arteries and lowest in the veins. In fact, if we list blood vessels in order according to the amount of blood pressure in them and draw a graph of this, as in Figure 8-13, the graph looks like a hill, with aortic blood pressure at the top and vena caval pressure at the bottom. This blood pressure hill is spoken of as the blood pressure gradient, or to be more exact, the term "blood pressure gradient" means the difference between two blood pressures. The blood pressure gradient for the entire systemic circulation is the difference between the average, or mean, blood pressure in the aorta and the blood pressure at the termination of the venae cavae where they join the right atrium of the heart. If you use the typical normal figures shown in Figure 8-13, the systemic blood pressure gradient figures out to be 100 millimeters of mercury pressure.

More important for you to remember than the amount of the blood pressure gradient is its function. The blood pressure gradient keeps blood flowing. When a blood pressure gradient is present, blood circulates. Conversely, when a blood pressure gradient is not present, blood does not circulate. For example, suppose that the blood pressure in the arteries were to decrease so that it became equal to the average pressure in arterioles. There would no longer be a blood pressure gradient between arteries and arterioles, and therefore there would no longer be a force to move blood out of arteries into arterioles. Circulation would stop, in other words, and very soon life itself would cease. This is why when arterial blood pressure is observed to be falling rapidly, whether in surgery or elsewhere in a hospital, emergency measures are quickly started to try to reverse this fatal trend.

What we have just said in the preceding paragraph may start you wondering why high blood pressure (meaning, of course, high arterial blood pressure) and low pressure are bad for circulation. High blood pressure is considered bad for several reasons. For one thing, if it becomes too high, it may cause the rupture of one or more blood vessels (for example, in the brain, as happens in a stroke). But low blood pressure also, as we have already seen, can be dangerous. If arterial pressure falls low enough, circulation and life cease. Massive hemorrhage, for instance, kills in this way.

How is blood pressure produced? What causes blood pressure, in other words, and what makes blood pressure change from time to time? To give a brief answer to these questions seems almost impossible, but we shall try to do so by discussing only the three factors that affect arterial blood pressure most directly. They are the volume of blood in the arteries, the beating of the heart, and the diameter of the arterioles. The volume of blood in the arteries directly determines the amount of arterial blood pressure. In general, the more blood in the arteries, the higher the blood pressure. Conversely, the less blood in the arteries the lower the blood pressure tends to be. Hemorrhage demonstrates well this relation between blood volume and blood pressure. In hemorrhage a marked loss of blood occurs, and this decrease in the volume of blood causes blood pressure to drop. In fact, the major sign of hemorrhage is a rapidly falling blood pressure.

Both the strength and the rate of the heartbeat affect blood pressure. Each time the left ventricle contracts, it squeezes a certain volume of blood into the aorta and on into other arteries. The stronger each contraction is, the more blood it pumps into the aorta and arteries. Conversely, the weaker each contraction is, the less blood it pumps. Suppose that one contraction of the left ventricle pumps 70 milliliters (about one-third of a cupful) of blood into the aorta, and suppose that the heart beats 70 times a minute. Seventy milliliters times 70 equals 4,900 milliliters. Almost five quarts of blood, in other words, would enter the aorta and arteries every minute. Now suppose that the heartbeat were to become weaker, and that each contraction of the left ventricle pumps only 50 milliliters of blood instead of 70 into the aorta. If the heart still contracts 70 times a minute, it will obviously pump much less blood into the aorta—only 3,500 milliliters instead of the more normal 4,900 milliliters per minute. This decrease in the heart's output of blood tends to decrease the volume of blood in the arteries, and the decreased arterial blood volume tends to decrease arterial blood pressure. In summary, the strength of the heartbeat affects blood pressure in this way—a stronger heartbeat tends to increase blood pressure and a weaker beat tends to decrease it.

Now let us turn our attention to the rate of the heartbeat and its effect on arterial blood pressure. We might expect that when the heart beats faster, more blood would enter the aorta and that therefore the arterial blood volume and blood pressure would increase. But this does not necessarily happen. Often when the heart beats faster, each contraction of the left ventricle takes place so rapidly that it squeezes out less blood than usual into the aorta. If, for example, the heart rate speeds up to 100 times a minute, each beat might pump only 40 milliliters of blood into the aorta instead of the normal 70 milliliters. What do you think would happen to blood pressure then? Knowing what you now do about the relation between the volume of blood pumped per minute by the heart, the volume of blood in the arteries, and the blood pressure, do you think that the blood pressure would increase, or decrease, or stay the same? Both arterial blood volume and blood pressure would tend to decrease under these conditions even though the heart rate had increased. What generalization, then, can we make? We can only say that an increase in the rate of the heartbeat tends to increase blood pressure and a decrease in the rate tends to decrease blood pressure. But whether a change in the heart rate actually produces a similar change in blood pressure, depends on whether the strength of the heart's beat also changes, and in which direction.

Another factor that we ought to mention in connection with blood pressure is the viscosity of blood, or in plainer language, its stickiness. If blood becomes less viscous than normal, blood pressure decreases. For example, if a person suffers a hemorrhage, fluid will move into his blood from his interstitial fluid. This dilutes the blood and decreases its viscosity, and blood pressure then falls because of the decreased viscosity. As you may know, after hemorrhage either whole blood or plasma is preferred to saline solution for transfusions. The reason is that saline solution is not a viscous liquid and so cannot keep blood pressure at a normal level.

Blood pressure does not stay the same every minute. It fluctuates even when we are healthy. For example, when we exercise strenuously, our blood pressure goes up. Not only is this normal but the increased blood pressure serves a good purpose. It tends to increase circulation, to bring more blood to muscles each minute, and thus to supply them with more oxygen and food for more energy.

A normal average arterial blood pressure is 120/80, or 120 millimeters of mercury systolic pressure (as the ventricles contract) and 80 millimeters of mercury diastolic pressure (as the ventricles relax).

Pulse

What you feel when you take a pulse is an artery expanding and then recoiling alternately. In order to feel a pulse you must place your fingertips over an artery that lies near the surface of the body and over a bone or other firm background. Six such places are given below. Try to feel your pulse at each of these locations. (You can locate all of them on Figure 8-8, except the temporal and facial arteries.)

1. *Radial artery*—at the wrist
2. *Temporal artery*—in front of the ear or above and to the outer side of the eye
3. *Common carotid artery*—in the neck along the front edge of the sternocleidomastoid muscle at the level of the lower margin of the thyroid cartilage
4. *Facial artery*—at the lower margin of the lower jaw bone on a line with the corners of the mouth
5. *Brachial artery*—at the bend of the elbow along the inner margin of the biceps muscle
6. *Dorsalis pedis artery*—on the front surface of the foot, just below the bend of the ankle joint

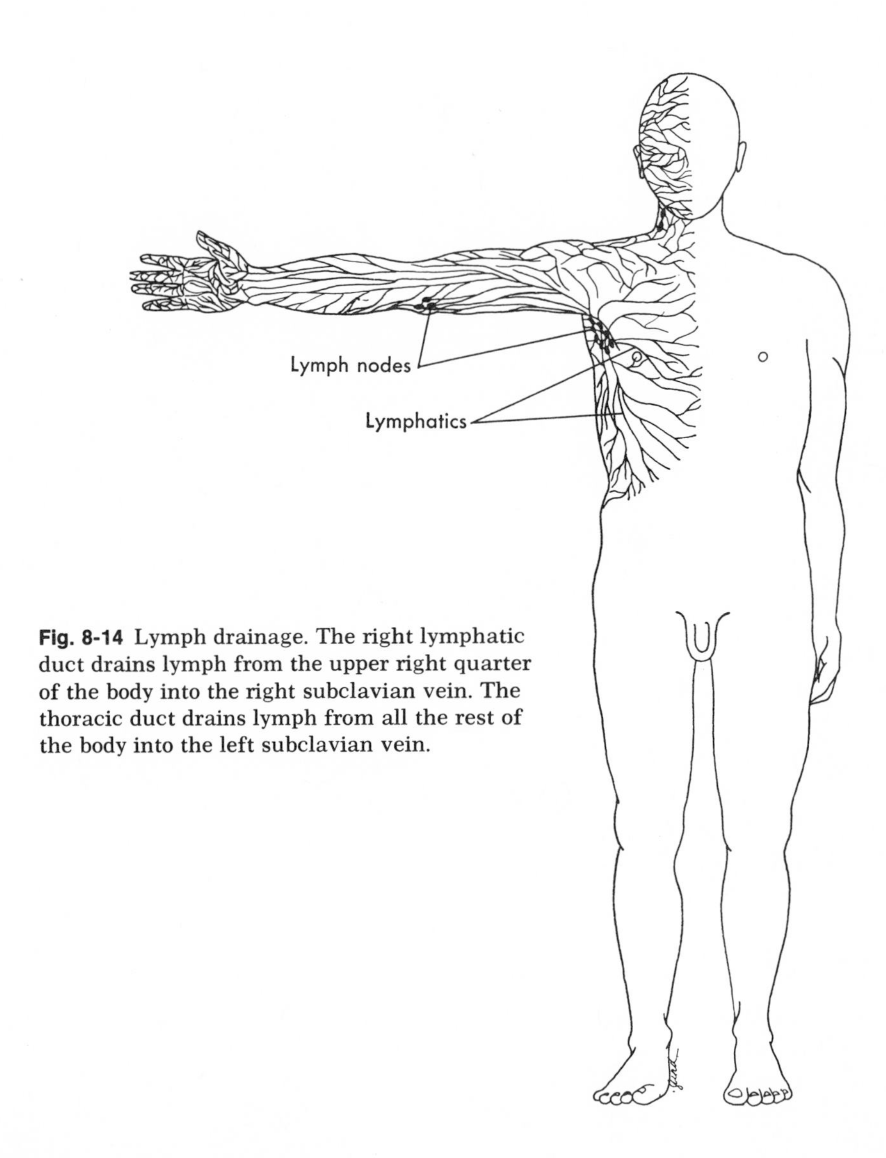

Fig. 8-14 Lymph drainage. The right lymphatic duct drains lymph from the upper right quarter of the body into the right subclavian vein. The thoracic duct drains lymph from all the rest of the body into the left subclavian vein.

Lymphatic system

The lymphatic system is not really a separate system of the body. It is part of the circulatory system since it consists of lymph, a moving fluid that comes from the blood and returns to the blood by way of the lymphatic vessels. Lymph forms in this way: blood plasma filters out of the capillaries into the microscopic spaces between tissue cells. Here the liquid is called *interstitial fluid,* or tissue fluid. Some of the interstitial fluid goes back into the blood by the same route it came out, that is, back through the capillary membrane. But most of the interstitial fluid enters tiny lymphatic capillaries to become lymph. It next moves on into larger lymphatics, and finally enters the blood in veins in the neck region. The largest lym-

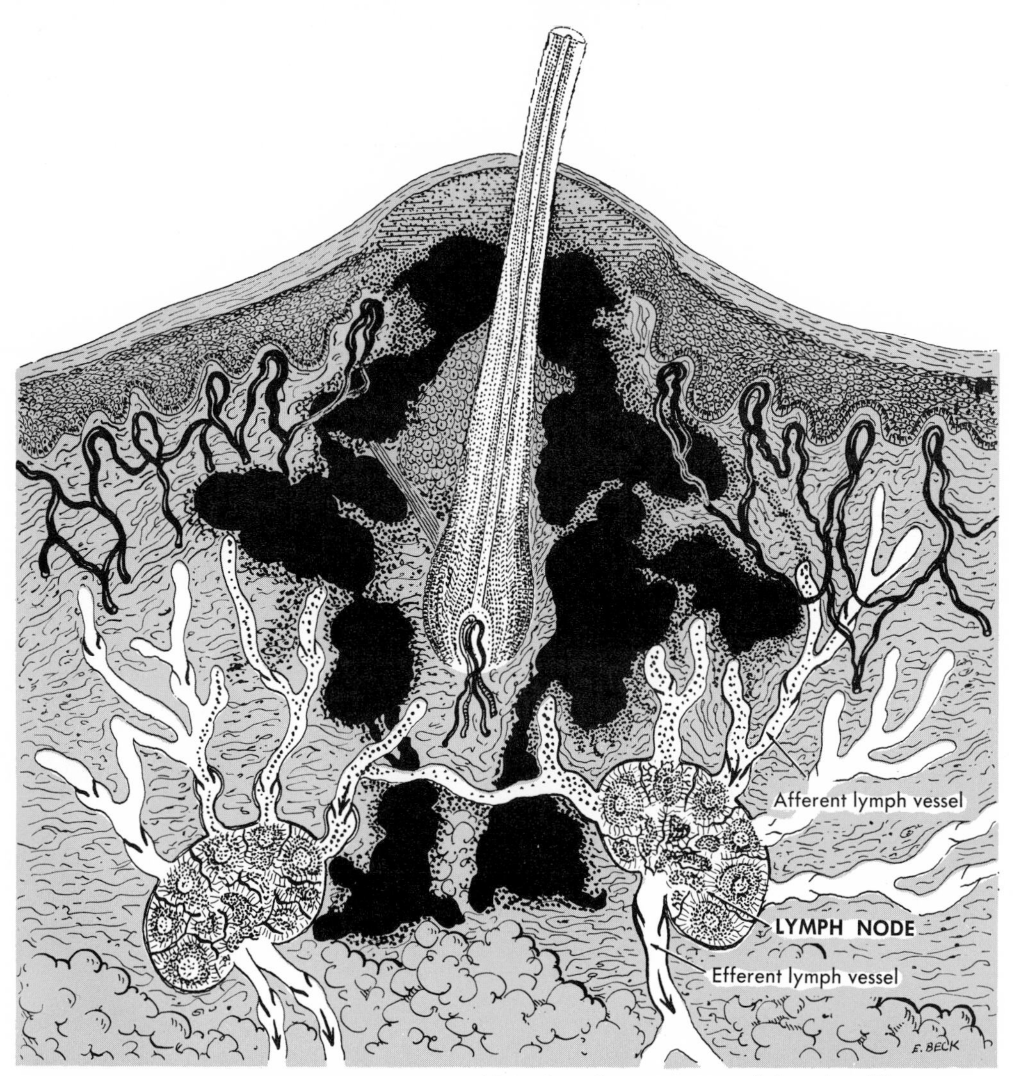

Fig. 8-15 Diagrammatic representation of lymph node structure and function. The drawing is of a skin section in which an infection surrounds a hair follicle. The black areas represent dead and dying cells (pus); black dots around the black areas represent bacteria. Leukocytes destroy many of the bacteria by phagocytosis, and as shown, many bacteria enter the lymph nodes by way of afferent lymphatics. The nodes filter and destroy most of the bacteria by enzymatic action.

phatic vessel is the *thoracic duct.* Lymph from about three quarters of the body (Figure 8-14) eventually drains into the thoracic duct and from it goes back into the blood. The thoracic duct joins the left subclavian vein at the angle where the internal jugular vein also joins it (Figure 8-9). Lymph from the rest of the body drains into the right lymphatic ducts, then on into the right subclavian vein.

Lymph nodes, or lymph glands as they are sometimes improperly called, are small oval structures located mainly in clusters along lymphatics (Figure 8-14). The structure of the lymph nodes makes it possible for them to perform an important protective function. They filter out injurious particles such as bacteria, soot, and cancer cells, thereby preventing them from entering the blood and circulating all over the body. Figure 8-15 illustrates lymph node structure and function. A surgeon uses his knowledge of lymph node function when he removes lymph nodes under the arms (axillary nodes) during an operation for breast cancer. These nodes may contain cancer cells filtered out of the lymph drained from the breast. A nurse, also, can use her knowledge of lymph node location and function. For example, when she takes care of a patient with an infected finger, she should watch the elbow and axillary regions for swelling and tenderness of the lymph nodes there; these nodes filter lymph returning from the hand and may become infected by the bacteria they trap. Lymph nodes also carry on another function, one mentioned earlier in this chapter. They form some of the white blood cells, namely lymphocytes and monocytes. (Lymphocyte functions were discussed on p. 116.)

Spleen

The spleen remains one of the mystery organs of the body. Quite a bit of doubt still exists about its functions. The three lower ribs provide a protective shelter over the spleen, which is located in the upper left corner of the abdominal cavity just under the diaphragm. Except for blood vessels and nerves, the spleen connects with no other organs.

The spleen's main functions seem to be to form lymphocytes and monocytes (as do the lymph nodes) and to act as the body's blood bank. The spleen can store almost two cups of blood and quickly release it back into circulation when more blood is needed—during strenuous exercise and after hemorrhage, for instance.

outline summary

BLOOD STRUCTURE AND FUNCTIONS

Cells

1 Kinds
  a Red cells (erythrocytes)
  b White cells (leukocytes)
    1 Neutrophils
    2 Eosinophils
    3 Basophils
    4 Lymphocytes
    5 Monocytes
2 Numbers
  a Red cells—$4\frac{1}{2}$ to 5 million per cubic millimeter of blood
  b White cells—5,000 to 9,000 per cubic millimeter of blood
3 Formation
  a Red bone marrow (myeloid tissue) forms all blood cells except lymphocytes and monocytes, which are formed by lymphatic tissue in lymph nodes and spleen
4 Functions
  a Red cells—transport oxygen and carbon dioxide
  b White cells—neutrophils and monocytes carry on phagocytosis; lymphocytes function to produce immunity
  c Platelets—release prothrombinase; substance that starts blood clotting

Blood plasma

1 Definition—blood minus its cells
2 Composition—water containing many dissolved substances; for example, foods, salts, hormones
3 Amount of blood—varies with size and sex; 4 to 6 quarts about average
4 Reaction—slightly alkaline

Blood types (or blood groups)

1 Type A blood—type A antigens in red cells; anti-B type antibodies in plasma
2 Type B blood—type B antigens in red cells; anti-A type antibodies in plasma
3 Type AB blood—type A and type B antigens in red cells; no anti-A or anti-B type antibodies in plasma; therefore, type AB blood is called universal recipient blood
4 Type O blood—no type A or type B antigens in red cells; therefore, type O blood is called universal donor blood; both anti-A and anti-B type antibodies in plasma
5 Rh positive blood—Rh factor antigen present in red cells
6 Rh negative blood—no Rh factor present in red cells; no anti-Rh antibodies present naturally in plasma; anti-Rh antibodies, however, appear in plasma of Rh negative person if Rh positive blood cells have been introduced into his body

Blood clotting

Summarized in Figure 8-2

HEART

1 Cavities – right atrium, right ventricle, left atrium, left ventricle
2 Wall – myocardium, composed of cardiac muscle
3 Lining – endocardium
4 Covering – pericardium
5 Valves – keep blood flowing in right direction through heart; prevent backflow
 a Tricuspid – at opening of right atrium into ventricle
 b Mitral (or bicuspid) – at opening of left atrium into ventricle
 c Pulmonary semilunars – at beginning of pulmonary artery
 d Aortic semilunars – at beginning of aorta
6 Conduction system
 a SA (sinoatrial) node – located in the wall of the right atrium near the opening of the superior vena cava
 b AV (atrioventricular) node – located in the right atrium along the lower part of the interatrial septum
 c AV (bundle of His) bundle – located in the septum of the heart
 d Purkinje fibers – located in the walls of the ventricles

BLOOD VESSELS

1 Kinds
 a Arteries – carry blood away from heart
 b Veins – carry blood toward heart
 c Capillaries – carry blood from arterioles to venules
2 Structure – see Figures 8-6 and 8-7
3 Names of main arteries – see Figure 8-8
4 Names of main veins – see Figure 8-9

CIRCULATION

1 Plan of circulation – see Figure 8-11
2 Portal circulation – detour of venous blood from stomach, pancreas, intestines, spleen through liver before return to heart
3 Plan of fetal circulation – see Figure 8-12

BLOOD PRESSURE

1 Blood pressure is push, or force, of blood in blood vessels
2 Highest in arteries, lowest in veins – see Figure 8-13
3 Blood pressure gradient causes blood to circulate – liquids can flow only from area where pressure is higher to where it is lower
4 Blood volume, heartbeat, and blood viscosity are main factors that produce blood pressure
5 Blood pressure varies within normal range from time to time

PULSE

1 Definition – alternate expansion and recoil of blood vessel wall
2 Places where you can count the pulse easily – see p. 131

LYMPHATIC SYSTEM

1 Consists of lymphatic vessels, lymph nodes, lymph
2 Lymph – the fluid in the lymphatic vessels; lymph comes from blood by plasma filtering out of blood capillaries to form interstitial fluid, some of which then enters the lymph capillaries to become lymph and be returned to blood by way of lymphatics; largest lymphatic is thoracic duct – drains lymph from all but upper right quarter of body into left subclavian vein
3 Lymph nodes
 a Located along certain lymphatics, usually in clusters; for example, at elbow, under arm, in groin, at knee
 b Functions – filter out injurious particles such as microorganisms and cancer cells from lymph before it returns to blood; form some white blood cells (lymphocytes and monocytes)

SPLEEN

1 Forms some white blood cells (lymphocytes and monocytes)
2 Serves as blood bank for body – stores blood until needed and then releases it back into circulation

new words

anemia
antibodies
antigens
diastolic pressure
embolism
fibrin
fibrinogen
leukemia
leukocytosis
leukopenia
myocarditis
pericarditis
plasma
prothrombin
serum
systolic pressure
thrombin
thrombosis
thrombus

review questions

1 What is phagocytosis? What cells perform this function?
2 What is blood plasma?
3 Suppose that you were asked what each of the following terms means: erythrocytes, leukemia, leukocytosis, and thrombocytes. What would you answer?
4 Suppose that your doctor told you that your "red count was 3 million." What does "red count" mean? Is 3 million normal? Might the doctor say that you had any of the following conditions – acidosis, anemia, leukopenia – with a red count of this amount? If so, which one?
5 If you had appendicitis or some other acute infection, would your white count be more likely to be 2,000, 7,000, or 15,000? Give a reason for your answer.
6 Your circulatory system is the transportation system of your body. Mention some of the substances it transports and tell whether each is carried in blood cells or in the blood plasma.
7 Do you think that advertisers' warnings to "guard against acid blood" are justified? Is there much danger of this?
8 Briefly explain what happens when blood clots, including what makes it start to clot.
9 You hear that a friend has a "coronary thrombosis." What does this mean to you?
10 Describe blood flow through the heart.
11 A patient has had an operation to repair the mitral valve. Where is this valve, and what is its function?
12 What are some differences between an artery, a vein, and a capillary?
13 Considering that the function of the circulatory system is to transport substances to and from the cells, do you think it is true that, in one sense, capillaries are our most important blood vessels? Give a reason for your answer.
14 The right ventricle of the heart pumps blood to and through only one organ. Which one?
15 What part of the heart pumps blood through the systemic circulation, that is, to and through all organs other than the lungs?
16 All blood returns from the systemic circulation to what part of the heart?
17 What part of the heart pumps blood through the pulmonary circulation, that is, to and through the lungs?
18 Blood returning from the pulmonary circulation (from the lungs, in other words) enters what part of the heart?
19 From which cavity of the heart does blood rich in oxygen leave the heart to be delivered to tissue capillaries all over the body?
20 How do arterial blood and venous blood differ with regard to their oxygen and carbon dioxide contents?
21 Does every artery carry arterial blood and every vein carry venous blood? If not, what exceptions are there?
22 Explain what is meant by "portal circulation."
23 Name the vein at the bend of the elbow into which substances are often injected and from which blood is sometimes withdrawn. (Check your answer with Figure 8-10.)
24 Nurses frequently have to take a patient's blood pressure. Why is blood pressure important?
25 Sometimes a woman's arm becomes very swollen for a while after removal of a breast and the nearby lymph nodes and lymphatics, including some of those in the upper arm. Can you think of any reason why swelling occurs?
26 You could live without your spleen since it does not do anything vital for the body. What functions does it perform?

9

The urinary system

The urinary system, as you might guess from its name, performs the functions of secreting urine and eliminating it from the body. What you might not guess so easily is how essential these functions are for healthy survival. Unless the urinary system operates normally, the normal composition of blood cannot long be maintained, and serious consequences soon follow. In this chapter we shall discuss the structure and function of each of the urinary system's organs. We shall also mention briefly some disease conditions produced by abnormal functioning of the urinary system.

Kidneys

Location and structure

There are two kidneys. They lie behind the abdominal organs against the muscles of the back. Usually the left kidney is a little larger than the right and a little farther above the waistline. A heavy cushion of fat normally encases each kidney and helps hold it in place; so in a very thin person, the kidneys may drop down a little (*renal ptosis*). This hinders urine drainage by putting a kink in the ureter, the tube that drains urine out of the kidneys.

The outer part of the kidney is called the *cortex*. (The word "cortex" comes from the Latin word that means bark or rind, so the cortex of an organ—the kidney, the brain, the adrenal glands—is its outer layer.) The interior part of the kidney, beneath the cortex, is the *medulla*.

Kidney cells and capillaries are arranged so that they form unique little structures called *nephrons*. Each nephron consists of three main parts: a glomerulus, a Bowman's capsule, and a tubule. The *glomerulus* is a network of blood capillaries tucked into the top of a microscopic-sized funnel-shaped structure. The top part of this structure is called *Bowman's capsule*, and the stem part is called the *renal tubule*. As you can see in Figure 9-2, different parts

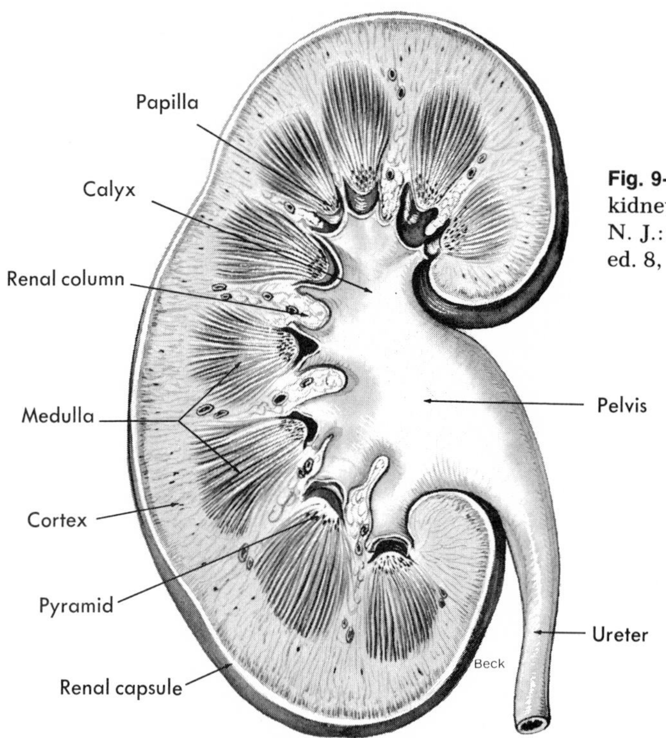

Fig. 9-1 Coronal section through the right kidney. (From Anthony, C. P., and Kolthoff, N. J.: Textbook of anatomy and physiology, ed. 8, St. Louis, 1971, The C. V. Mosby Co.)

of the renal tubule have different names. The first segment of each tubule is called the proximal convoluted tubule—proximal because it lies nearest the tubule's origin from Bowman's capsule, and convoluted because it twists around to form several coils. The proximal convoluted tubule becomes the descending limb and then the ascending limb of the loop of Henle. The ascending limb becomes the distal convoluted tubule, which terminates in a straight, or collecting, tubule that opens into the renal pelvis.

Functions

The structure of nephrons makes them able to carry on their function of urine formation. Because the walls of the glomerular capillaries and of Bowman's capsules are very thin membranes, water and dissolved substances (except albumin and other blood proteins) filter rapidly out of the blood in the glomeruli into Bowman's capsules. This filtrate then trickles down the convoluted tubules; and as it does so, a large part of the water goes back into the blood, that is, is reabsorbed into capillaries around the tubules. Dissolved substances (solutes) also leave the tubule filtrate to return to the blood. Glucose, for instance, is entirely reabsorbed, so that none of it is wasted by being lost in the urine. Exceptions to this normal rule, however, do occur. For example, in diabetes mellitus, if blood glucose concentration increases above a certain level, the tubular filtrate then contains more glucose than the kidney tubule cells can reabsorb into the blood. Some of the glucose, therefore, remains behind in the urine; sugar in the

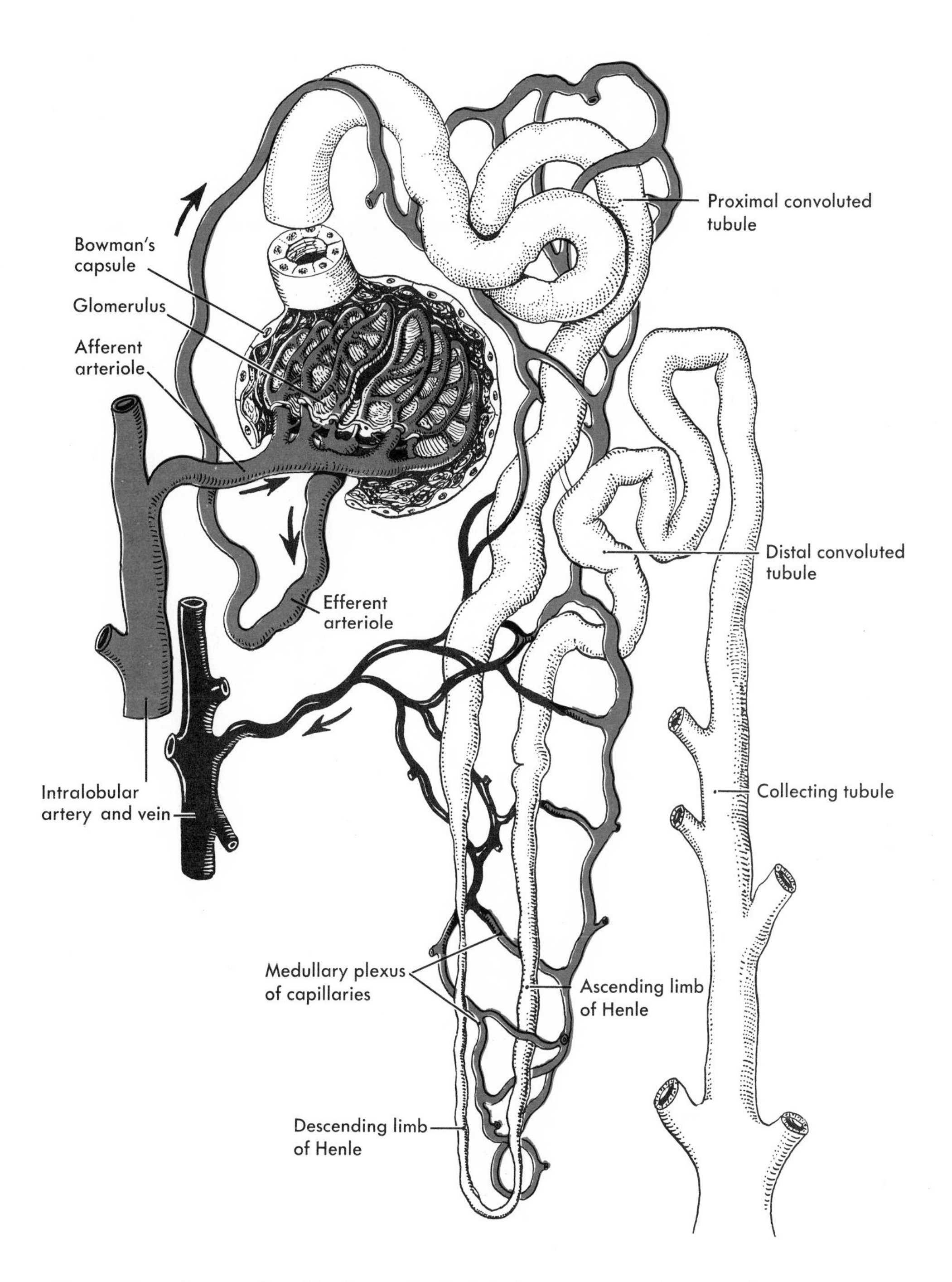

Fig. 9-2 A nephron and its blood vessels. Each kidney contains about a million nephrons. Arrows indicate the direction of blood flow: From the intralobular artery → the afferent arteriole → the glomerulus → the efferent arteriole → the peritubular capillaries (around the tubules) → the venules (shown in black) → the intralobular vein.

urine (*glycosuria* or *glucosuria*) is a common sign of diabetes.

Like glucose, sodium chloride and other salts filter out of glomerular blood. Unlike glucose, however, salts are only partially reabsorbed from the tubule filtrate. The amount reabsorbed varies from time to time. When, for example, you salt food heavily, your kidneys reabsorb less salt than when you use salt sparingly. Thus, your body gets rid of excess salt by way of the urine, and in this way tends to keep the blood's salt concentration normal. This is an extremely important matter because cells are damaged by either too much or too little salt in the fluids around them.

To summarize, we can say that three processes in succession accomplish the task of urine formation (Figure 9-3):

1. *Filtration*—of water and dissolved substances out of the blood in the glomeruli into Bowman's capsules
2. *Reabsorption*—of water and dissolved substances out of the kidney tubules back into the blood (Note that this process prevents substances needed by the body from being lost in the urine. Usually 97% to 99% of the water filtered out of the glomerular blood is retrieved from the tubules.)
3. *Secretion*—of hydrogen ions (H^+), potassium ions (K^+), and certain drugs (for example, penicillin)

Secretion, like reabsorption, is also carried on by kidney tubules. Substances secreted move out of blood into tubule urine—a kind of reabsorption in reverse, since substances reabsorbed move out of tubule urine into blood. Kidney tubule secretion is an essential process. Kidney tubules must secrete varying amounts of the acid hydrogen ions in order to maintain the slightly alkaline reaction of blood that is essential for healthy survival.

The body has ways to control both the amount and the composition of the urine that it secretes. It does this mainly by controlling the amount of water and dissolved substances reabsorbed by the convoluted tubules. A hormone (antidiuretic hormone, or ADH) from the posterior pituitary gland, for instance, tends to decrease the amount of urine by causing tubule cells to reabsorb more water. As a result, less water is lost from the body or more water is retained—whichever way you wish to say it. At any

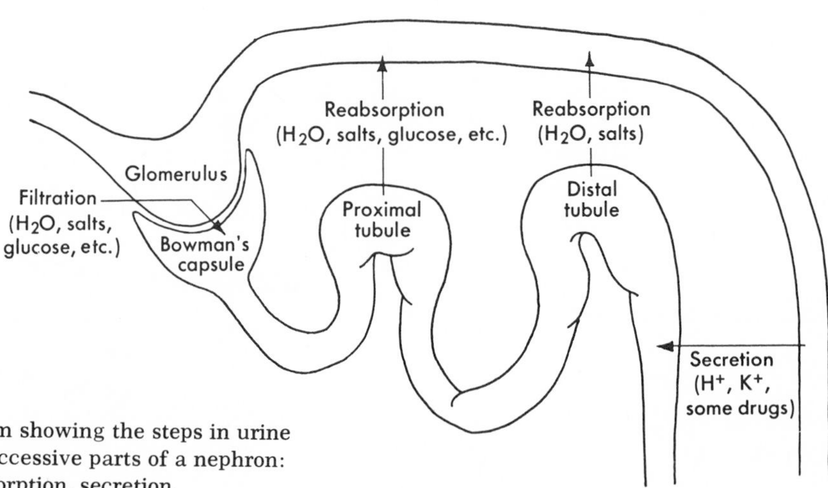

Fig. 9-3 Diagram showing the steps in urine formation in successive parts of a nephron: filtration, reabsorption, secretion.

rate, for this reason ADH is accurately described as the "water-retaining hormone." You might also think of it as the "urine-decreasing hormone." The hormone aldosterone, secreted by the adrenal cortex, plays an important part in controlling the kidney tubules' reabsorption of salt. Primarily it stimulates the tubules to reabsorb sodium salts at a faster rate. Secondarily, aldosterone tends also to increase tubular water reabsorption. The term "salt- and water-retaining hormone," therefore, is a descriptive nickname for aldosterone.

Sometimes the kidneys do not excrete normal amounts of urine—as a result of kidney disease, cardiovascular disease, or stress. (See question 7, p. 142.) Here are some terms associated with abnormal amounts of urine:

1. *Anuria*—literally, absence of urine
2. *Oliguria*—scanty urine
3. *Polyuria*—unusually large amounts of urine

Ureters

Urine drains out of the collecting tubules of each kidney into the renal pelvis and on down the ureter into the urinary bladder (Figure 9-1). The *renal pelvis* is the basin-like upper end of the ureter and is located inside the kidney. Ureters are narrow tubes less than 1/4 inch wide but 10 to 12 inches long. Mucous membrane lines both ureters and renal pelves. Smooth muscle fibers in the ureter walls contract to produce a peristaltic movement that forces urine down into the bladder.

Urinary bladder

Elastic fibers and involuntary muscle fibers in the wall of the urinary bladder make it well suited for its functions of expanding to hold variable amounts of urine and then contracting to empty itself. Most people feel the desire to void when the bladder contains about 1/2 pint (250 milliliters) of urine. Mucous membrane lines the urinary bladder. It lies in folds (rugae) when the bladder is empty.

Urinary *retention* is a condition in which no urine is voided. The kidneys secrete urine, but the bladder for one reason or another cannot empty itself. In urinary *suppression*, the opposite is true. The kidneys do not secrete any urine, but the bladder retains the ability to empty itself. Do you think anuria could be a symptom of either suppression or retention?

Incontinence is a condition in which the patient voids urine involuntarily. It frequently occurs in patients who have suffered a stroke or spinal cord injury.

Urethra

To leave the body, urine passes from the bladder, down the urethra, and out its external opening, the *urinary meatus*. In other words, the urethra is the lowermost part of the urinary tract. The same sheet of mucous membrane that lines the renal pelves, ureters, and bladder extends down into the urethra too—an interesting structural feature because it accounts for the fact that an infection of the urethra may spread upward throughout the urinary tract. The urethra is a narrow tube. It is only about 1 1/2 inches long in a woman but about 8 inches long in a man.

Diseases of the urinary system

Inflammation of the kidney—*nephritis*—occurs relatively often. If the inflammation affects the glomeruli primarily, the disease is called *glomerulonephritis*. If it affects the renal pelvis as well as the kidney, it is called *pyelonephritis.*

Nephrosis is a disease in which degenerative changes develop in the renal tubules.

Cystitis is an inflammation of the mucous lining of the urinary bladder.

outline summary

KIDNEYS

Structure

1 Cortex – outer layer of kidney
2 Medulla – interior part of kidney
3 Nephrons – microscopic functional units of kidneys; each consists of a glomerulus, a Bowman's capsule, and a renal tubule

Function

1 Form urine by means of glomerular filtration, tubule reabsorption, and tubule secretion
2 Amount of urine controlled mainly by hormones – ADH from posterior pituitary gland and aldosterone from adrenal cortex

URETERS

1 Structure
 a Narrow long tubes with expanded upper end (renal pelvis) located inside kidney
 b Mucous lining
2 Function – drain urine from kidneys to urinary bladder

URINARY BLADDER

1 Structure
 a Elastic muscular organ, capable of great expansion
 b Lined with mucous membrane arranged in rugae, like stomach mucosa
2 Functions
 a Store urine before voiding
 b Voiding

URETHRA

1 Structure
 a Narrow short tube from urinary bladder to exterior
 b Lined with mucous membrane
 c Opening of urethra to exterior called urinary meatus
2 Function
 a Serves as passageway by which urine leaves bladder for exterior
 b Passageway from which male reproductive fluid leaves body

new words

anuria
Bowman's capsule
cystitis
glomerulonephritis
glycosuria
nephritis
nephrosis
oliguria
polyuria
pyelonephritis
renal ptosis
urinary retention
urinary suppression

review questions

1 What organs form the urinary system?
2 What and where are the glomeruli and Bowman's capsules?
3 Explain briefly the functions of the glomeruli and Bowman's capsules
4 Explain briefly the function of the renal tubules.
5 What kind of membrane lines the urinary tract?
6 Explain briefly the function of ADH. What is the full name of this hormone? What gland secretes it?
7 Suppose that ADH secretion increases markedly. Would this increase or decrease urine volume? Why?
8 What hormone might appropriately be nicknamed "the water-retaining hormone"?
9 What hormone might appropriately be nicknamed "the salt- and water-retaining hormone"?
10 What is the urinary meatus?
11 What and where are the ureters and the urethra?
12 Define briefly each term listed under "new words."

unit five

Systems that reproduce the body

10

The male and female reproductive systems

"Fearfully and wonderfully made" we truly are. Almost any one of the body's structures or functions might have inspired this statement, but of them all perhaps the reproductive systems best deserve such praise. Their achievement? The miracle of duplicating the human body. Their goal? The survival of the human species.

As you probably noticed, we used the plural, reproductive systems, in the preceding paragraph and in the chapter title. The male reproductive system consists of one group of organs and the female reproductive system consists of another group. These two systems differ in structure, but they share a common function—that of reproducing the human body. We shall discuss the male reproductive system first and then the female reproductive system.

Male reproductive system

Structural plan

So many organs make up the male reproductive system that we need to look first at the structural plan of the system as a whole. It consists of a pair of main sex glands, a series of ducts from these to the exterior, accessory sex glands, and the external reproductive organs. Table 10-1 lists the names of all of these structures, and Figure 10-1 shows the location of most of them.

External genitals

Male external reproductive organs (*genitals*) consist of two organs, the scrotum

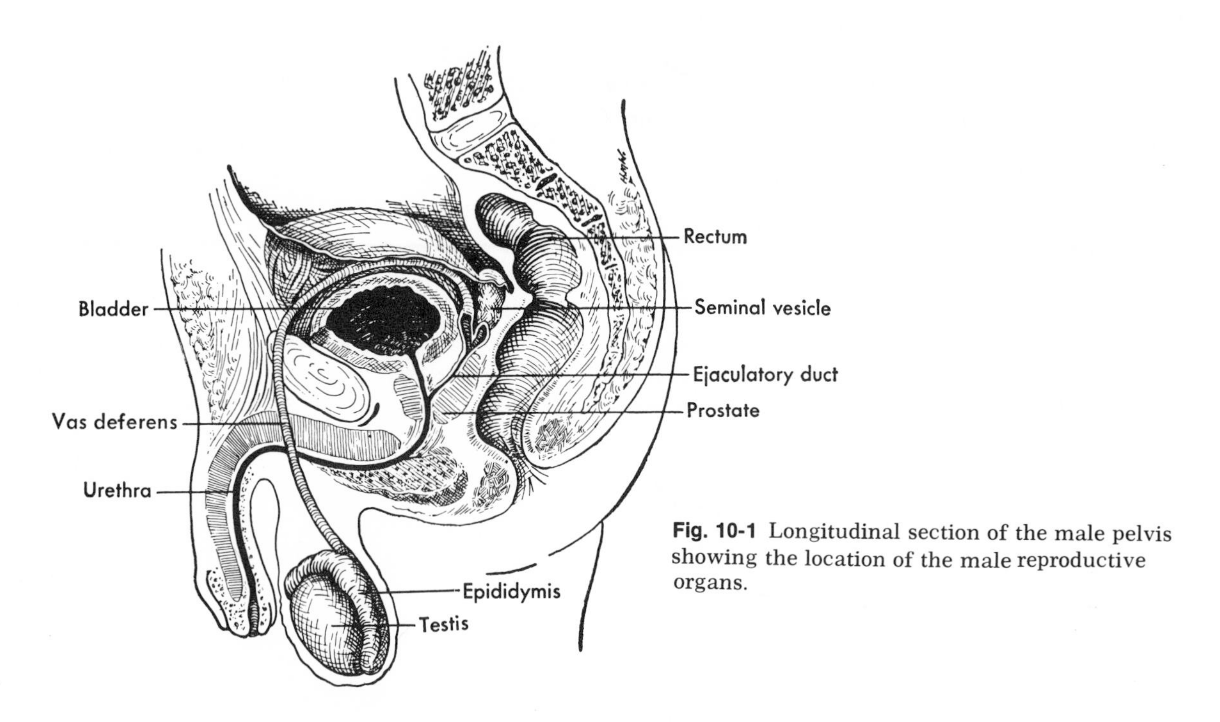

Fig. 10-1 Longitudinal section of the male pelvis showing the location of the male reproductive organs.

Table 10-1. Male reproductive organs

Main sex gland	Testes (right testis and left testis)
Ducts	Epididymis, vas deferens, ejaculatory duct (two of each), and one urethra
Accessory sex glands	Seminal vesicle, bulbourethral gland (two of each), and one prostate gland
External genitals	Scrotum and penis

and the penis. A special kind of tissue known as erectile tissue composes most of the interior of the penis. Erectile tissue contains many small spaces that usually are collapsed. Under the stimulus of the sexual emotion, blood floods these spaces, distending them enough to cause enlargement and rigidity of the organ.

Skin covers the outside of the penis. At the lower end of this organ, the skin is folded over double to form a somewhat loose-fitting casing called the foreskin (or prepuce). Circumcision involves cutting the foreskin so that it will not fit too tightly and cause irritation.

Testes

To judge the testes by their size would be to underestimate their importance. Each of these two oval-shaped glands is only about 1½ inches long and 1 inch wide, yet each testis forms millions of male sex cells (*spermatozoa*, or *sperm*), any one of which may join with a female sex cell (*ovum*) to become a new human being. Each testis also secretes the male sex hormone testosterone; the hormone that in a few short months transforms a little boy into a man. Testosterone lowers the pitch of his voice; it makes his muscles grow large and strong; it even changes the size and shape of his bones.

The testes are located in an external organ—a pouchlike, skin-covered structure called the *scrotum*. Each testis consists of several sections (lobules), and each lobule consists of a narrow but long and coiled *seminiferous tubule*. From the age of puberty on, the seminiferous tubules are almost continuously forming spermatozoa or sperm. *Spermatogenesis* is the name of this process. The other function of the testes is to secrete the male hormone testosterone. This function, however, is carried on by the *interstitial cells* of the testes, not by its seminiferous tubules. Numerous clusters of interstitial cells are located in the connective tissue between the seminiferous tubules.

Testosterone serves the following general functions:

1. It masculinizes. The various characteristics that we think of as "male" develop because of testosterone's influence. For instance, when a young boy's voice changes, it is testosterone that brings this about.

2. It promotes and maintains the development of the male accessory organs (prostate gland, seminal vesicles, and so on).

3. It has a stimulating effect on protein anabolism. Testosterone thus is responsible for the greater muscular development and strength of the male. A good way to remember testosterone's functions is to think of it as "the masculinizing hormone" and "the anabolic hormone."

Control of testosterone secretion

How much testosterone the interstitial cells of the testes secrete depends upon how much of another hormone—the interstitial-cell-stimulating hormone (ICSH)—the anterior pituitary gland secretes. The more ICSH secreted, the more it stimulates the interstitial cells to secrete more testosterone. Or we can say this a shorter way: a high blood concentration of ICSH stimulates testosterone secretion. The resulting high blood concentration of testosterone then feeds back via the circulating blood to influence ICSH secretion by the anterior pituitary gland. But here the effect is quite different. Testosterone exerts a negative effect on ICSH secretion. A high blood concentration of testosterone inhibits ICSH secretion instead of stimulating it. This is an example of a "negative feedback control mechanism"—a term used frequently in our computer-conscious world.

Epididymis, vas deferens, ejaculatory duct, and urethra

The role of the vas deferens in birth control has recently become very important. To find out why, we need first to trace the route by which sperm leave the male reproductive tract in order to enter the female tract. As you read the description of this route in the next sentences, follow it in Figure 10-1. Sperm are formed in the testes by the seminiferous tubules. From the seminiferous tubules, sperm by the millions stream into a narrow but long and tightly coiled duct, the *epididymis*. From the duct of the epididymis, sperm continue on their way through one of the vas deferens into an ejaculatory duct from which they move down the urethra and out of the body. Note in Figure 10-1 the location of the vas deferens in the scrotum. Here these small tubes lie near the surface. This fact makes it possible for a surgeon to make a small incision in the scrotum and quickly and easily cut out a section of each vas and tie off each of its separated ends. The technical name for this minor surgery is a bilateral partial *vasectomy*. Although a man's seminiferous tubules may continue to form sperm after he has had a vasectomy, they can no longer leave his body.

A part of their exit route has been cut away and no detour route has been provided. Therefore, the man has become sterile, that is, he can no longer father children. (Reversing this operation, however, is possible in some cases, if at a later date the man wishes to be fertile again.) A vasectomy does not change a man's ability to have an erection or an ejaculation. As explained in the next paragraph, structures that lie beyond the vas deferens produce the semen.

Seminal vesicles, prostate gland, and bulbourethral glands

The two seminal vesicles, one prostate gland, and two bulbourethral glands are accessory male glands that produce alkaline secretions. These secretions constitute the gelatinous fluid part of the *semen.* Normally 3 to 5 milliliters (about 1 teaspoonful) of semen is ejaculated at one time, and each milliliter contains over 60,000,000 sperm. After a successful bilateral vasectomy, about the same amount of semen may be ejaculated as before, but it contains no sperm. In short, a vasectomy makes a man sterile but not impotent. It constitutes one of the many modern methods of birth control.

The *prostate gland* claims importance not so much for its function as for its trouble making. In older men it often becomes inflamed and enlarged, squeezing on the urethra, which runs through the center of the doughnut-shaped prostate. Sometimes, in fact, the prostate enlarges so much that it closes off the urethra completely. Urination then becomes impossible. (Should you refer to this as urinary retention or urinary suppression? If you are not sure, check your answer on p. 141.)

The small *bulbourethral* (Cowper's) *glands* lie one on either side of the urethra just below the prostate gland. Like the seminal vesicles and the prostate, the bulbourethral glands add an alkaline secretion to the semen. Sperm survive and remain fertile longer in an alkaline fluid than in an acid one.

Female reproductive system

Structural plan

The structural plan of the female reproductive system resembles that of the male system. Like the male reproductive system, the female reproductive system consists of a pair of main sex glands, ducts from these to the exterior, accessory sex glands, and external genitals. To find out the names of these structures in the female, consult Table 10-2.

External genitals

The scientific name for the female external genitals is the *vulva.* Identify its various parts in Figure 10-3.

Ovaries

Although male and female reproductive systems resemble each other in plan, they differ from one another in details. For example, the testes, the main sex glands of the male, are not located inside a body cavity but lie in an external skin-covered pouch, the scrotum. In the female, however, the main sex glands, the ovaries, lie within the pelvic cavity. Also, there are two differences between male and female reproductive systems that sometimes assume clinical importance. One is that the ducts for the female sex glands, the ovaries, do not attach directly to the ovaries. Later we shall explain the importance of this fact (p. 150). The other clinically significant difference is that the male urethra serves as the outlet for both the urinary tract and the reproductive tract. This is not the case in the female. In a woman's

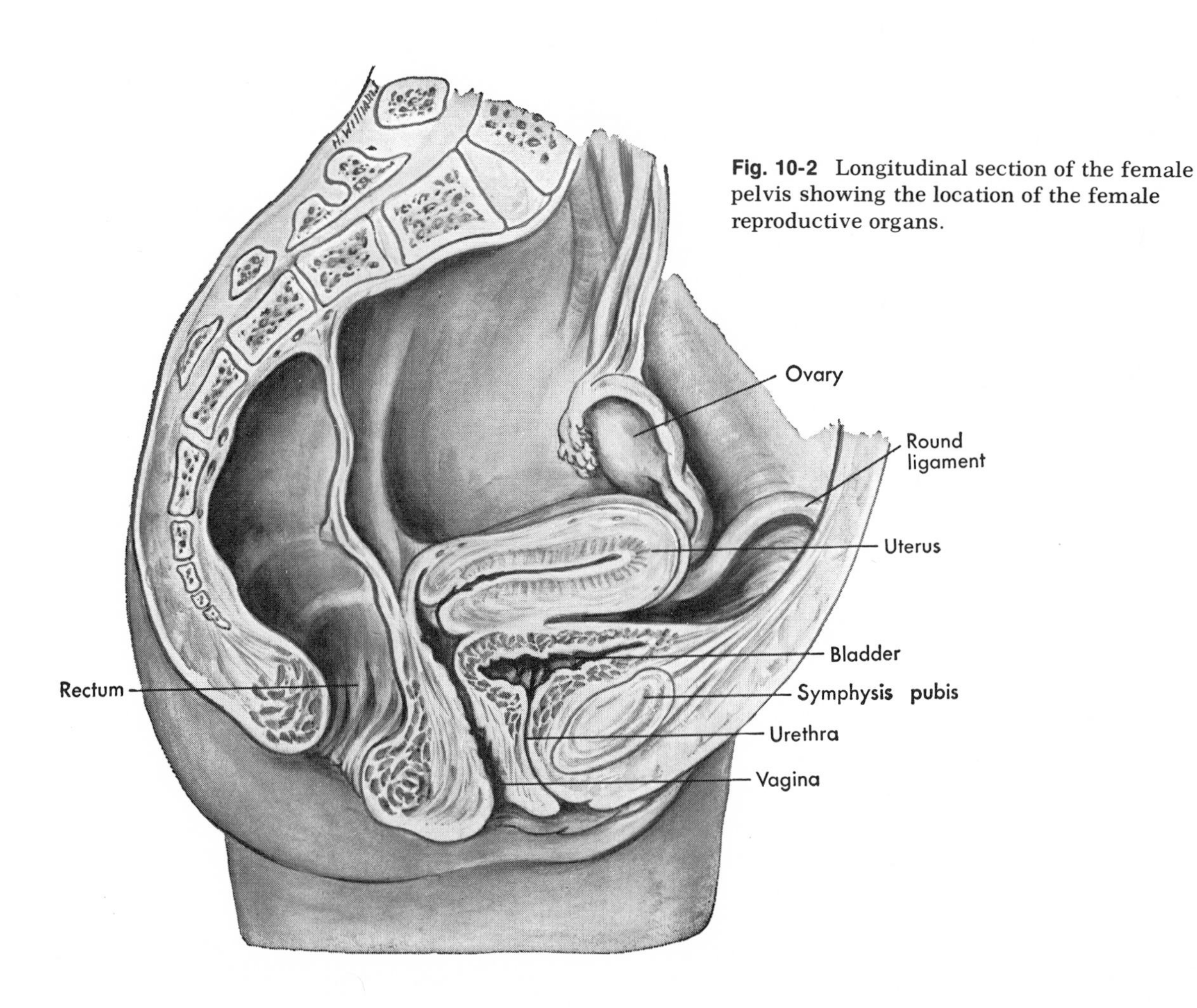

Fig. 10-2 Longitudinal section of the female pelvis showing the location of the female reproductive organs.

Table 10-2. The female reproductive system

Main sex gland	Ovaries
Ducts	Uterine tubes, uterus, and vagina
Accessory glands	Bartholin's glands
External genitals	Vulva (or pudendum)

body, the urethra serves only the urinary tract and a separate tube, the vagina, serves the reproductive tract. This difference can be important. For example, if a man contracts gonorrhea, the infection can spread from the urethra throughout both his reproductive and urinary tracts.

Several thousand sacs, too small to be seen without a microscope, make up the bulk of each ovary. They are called *graafian follicles,* in honor of a Dutch anatomist who discovered them some 300 years ago. Within each follicle lies an immature *ovum,* the female sex cell.

Like the testes, the ovaries produce both

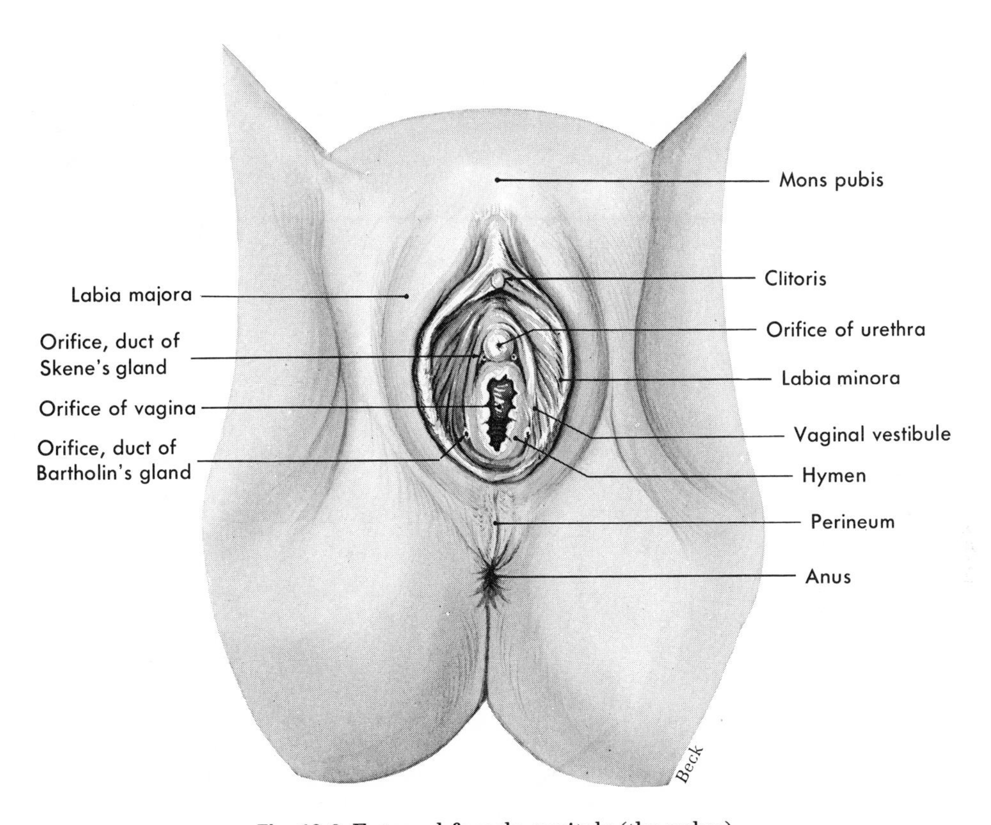

Fig. 10-3 External female genitals (the vulva).

sex cells and sex hormones. However, the ovaries usually produce only one mature sex cell—mature ovum, that is—at a time, whereas the testes produce many millions of mature sperm. The ovaries secrete two kinds of female hormones, namely, estrogens and progesterone. The testes on the other hand, secrete only one kind of male hormone, namely, *androgens;* testosterone is the only important androgen. The only endocrine gland cells of the testes, as you read on p. 146, are the interstitial cells. But the ovaries contain two endocrine glands, namely, the graafian follicles, and the *corpus luteum.* Graafian follicles secrete one kind of hormone, estrogens, whereas the corpus luteum secretes chiefly progesterone and also some estrogens.

Just as androgens are the masculinizing hormone, so estrogens are the feminizing hormone. To discover estrogens' two chief functions, see Figure 10-4. Look next at Figure 10-5; it will tell you the two chief functions progesterone performs.

Uterine tubes (fallopian tubes)

The uterine tubes serve as ducts for the ovaries even though they are not attached to them. The outer end of each tube curves over the top of each ovary. The inner end

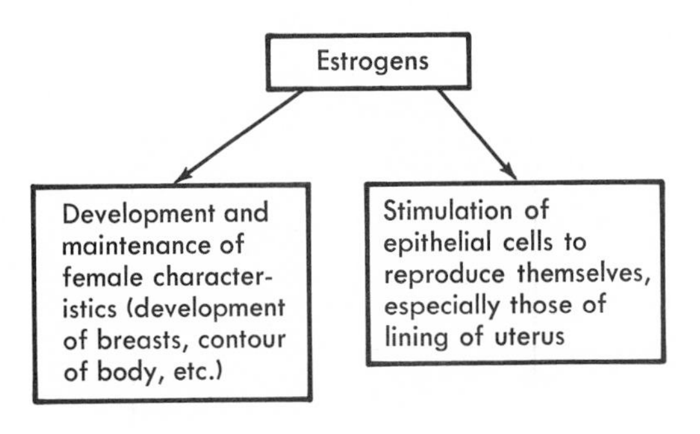

Fig. 10-4

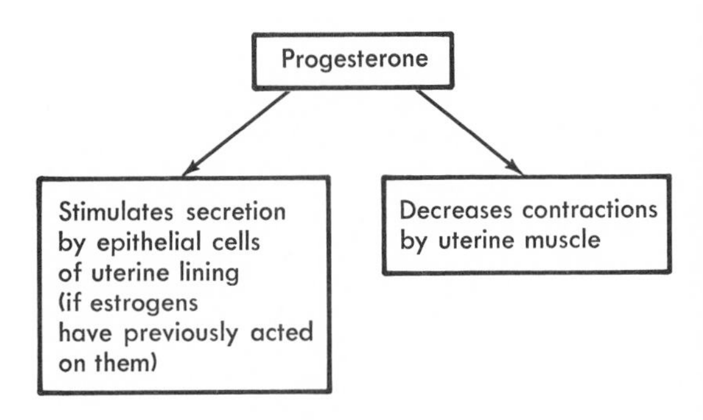

Fig. 10-5

of each uterine tube attaches to the uterus, and the cavity inside the tube opens into the cavity in the uterus. Each tube measures about 4 inches in length.

After ovulation, the ovum finds its way into one of the uterine tubes. This is where fertilization (union of a sperm with an ovum) normally occurs. Occasionally, however, because the tubes are not actually connected to the ovaries, an ovum does not enter a tube but becomes fertilized in the pelvic cavity. The term *ectopic pregnancy* means a pregnancy that develops outside of its proper place in the cavity of the uterus.

Uterus

The uterus is a small organ—only about the size of a pear—but it is extremely strong. It is almost all muscle with only a small cavity inside. During pregnancy the uterus grows many times larger so that it becomes big enough to hold a baby plus a considerable amount of fluid. The uterus is composed of two parts: an upper portion, the *body*, and a lower narrow section, the *cervix*. Just above the level where the uterine tubes attach to the body of the uterus, it rounds out to form a bulging prominence called the *fundus*. Except during pregnancy, the uterus lies in the pelvic cavity just behind the urinary bladder. By the end of pregnancy it becomes large enough to extend up to the top of the abdominal cavity. It then presses against the underside of the diaphragm—a fact that explains such a comment as "I can't seem to take a deep breath since I've gotten so big," made by many women late in their pregnancies.

The uterus functions in three processes—menstruation, pregnancy, and labor. The corpus luteum stops secreting progesterone and slows down on the secretion of estrogens about twelve days after ovulation. Two days later, when the blood progesterone and estrogen concentrations are at their lowest, menstruation starts. Bits of *endometrium* (mucous membrane lining of the uterus) pull loose, leaving torn blood vessels underneath. Blood and bits of endometrium trickle out of the uterus into the vagina and out of the body. Immediately after menstruation the endometrium starts to repair itself. It again grows thick and becomes lavishly supplied with blood in preparation for pregnancy. But if fertilization does not take place, the uterus once more sheds the lining made ready for a pregnancy that did not occur. Because these changes in the uterine lining continue to repeat themselves over and over, they are spoken of as the *menstrual cycle*. For a description of this cycle in the form of a diagram, see Figure 10-8.

Menstruation first occurs at puberty, of-

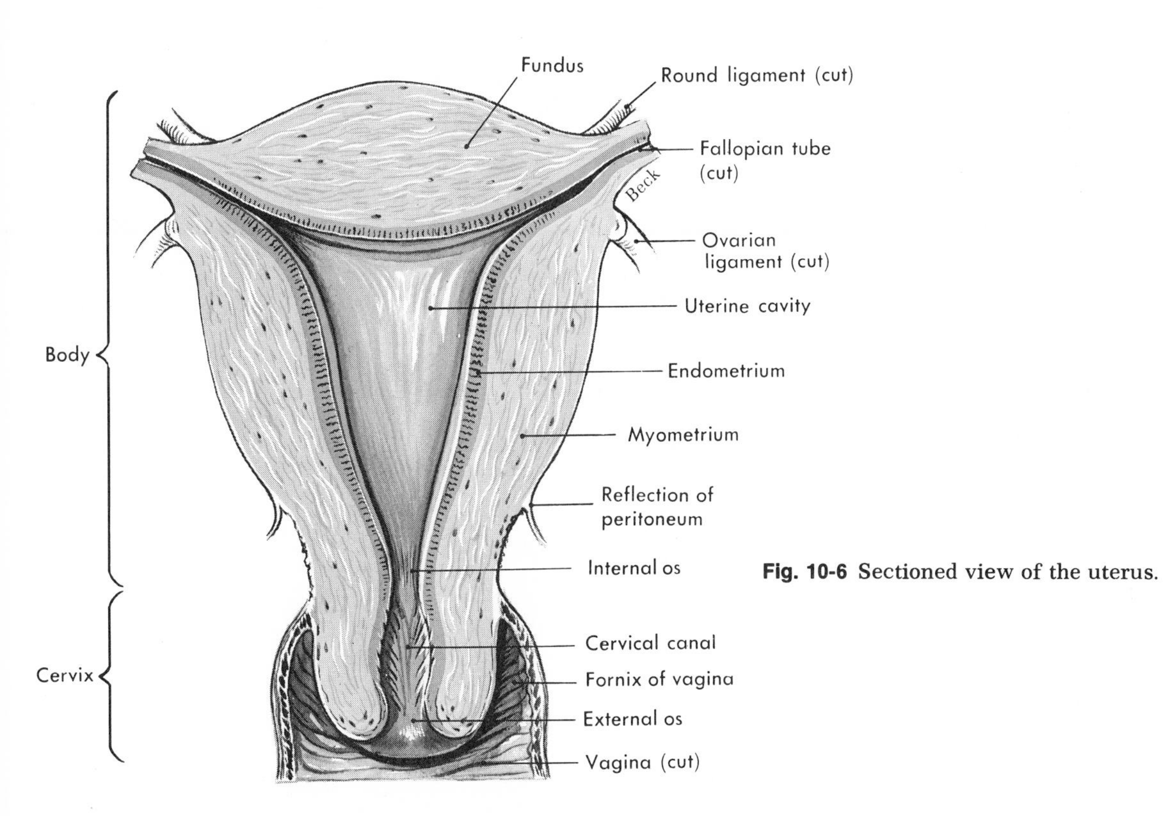

Fig. 10-6 Sectioned view of the uterus.

ten around the age of 13 years. Normally, it repeats itself about every 28 days or thirteen times a year for some 30 years or so before it ceases (*menopause*, or *climacteric*), when a woman is somewhere around the age of 45 years.

Vagina

The vagina is a distensible tube made mainly of smooth muscle and lined with mucous membrane. It lies in the pelvic cavity between the urinary bladder and the rectum. As the part of the female reproductive tract that opens to the exterior, the vagina is the organ that sperm enter on their journey to meet an ovum, and it is also the organ from which a baby emerges to meet its new world.

Bartholin's glands

One of the small Bartholin's glands lies to the right of the vaginal outlet and one to the left of it. Secreting a lubricating fluid is the function of Bartholin's glands. Their ducts open into the space between the labia minora and the hymen (Figure 10-3). Bartholinitis, an infection of these glands, occurs frequently. It often develops, for example, when a woman contracts gonorrhea.

Breasts

Each breast consists of fifteen to twenty-five divisions or lobes that are arranged radially. Each lobe consists of several lobules, and each lobule consists of secreting cells. Grapelike clusters of the secreting cells surround small ducts. The small ducts unite so that only one duct leads from each lobe to an opening in the nipple. The colored area around the nipple is the *areola*. It changes from a delicate pink to brown early in pregnancy and never quite returns to its original color.

Menstrual cycle

The menstrual cycle consists of a great many changes—in the uterus, ovaries, vagina, and breasts, and in the functioning of the anterior pituitary gland. In the majority of women these changes occur with almost precise regularity throughout their reproductive years. The first indication of changes comes with the event of the first menstrual period. (*Menarche* is the scientific name for the beginning of the menses.)

A typical menstrual cycle covers a period of 28 days. Each cycle consists of three

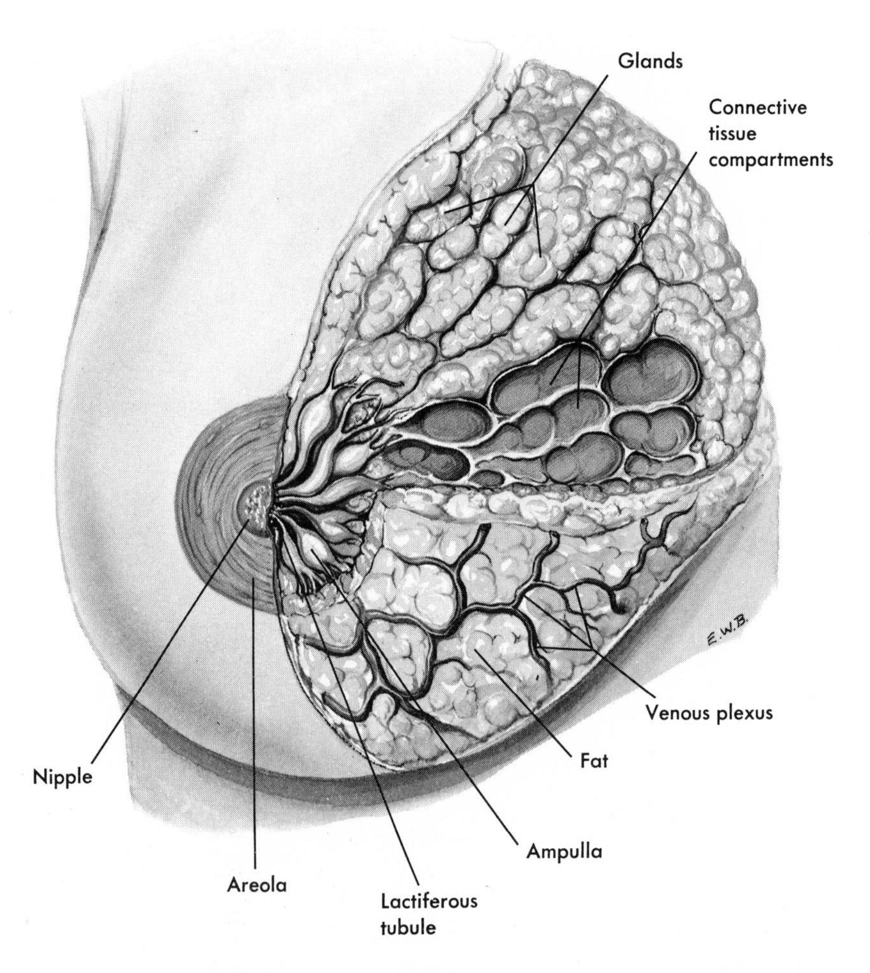

Fig. 10-7 Anterior view of the breast. The skin has been removed in the lower right quadrant to reveal the venous plexus overlying the adipose (fatty) tissue. In the upper right quadrant the adipose tissue has been removed to show the alveoli of the glands. The glands have been removed in a small area to reveal the connective tissue compartments that separate the lobules.

phases. Although they have been called by several different names, we shall call them the menses, the postmenstrual phase, and the premenstrual phase. Now examine Figure 10-9 to find out what changes take place during each phase of the menstrual cycle. Be sure you do not overlook the event that occurs on day 14 of a 28-day cycle.

Physiologists today know a great deal about the functioning of the female reproductive system. They also are continuing to learn more about it each year. They now know, for example, that many changes take place in a woman's body each month. They know when these changes occur, and they also know at least in part what causes them. All of this has become enormously important knowledge, since overpopulation looms so large as a threat to man's healthy survival on this planet. Knowledge about the body's method of controlling the events of the menstrual cycle has made possible most of our modern methods of birth control. Knowledge about the male reproductive system has lead to other methods—vasectomy, for example. Current and proposed research gives promise of still other methods that may prove even more practical.

From the first to about the seventh day of the menstrual cycle, the anterior pituitary gland secretes increasing amounts of FSH (follicle-stimulating hormone). A high blood concentration of FSH stimulates several immature ovarian follicles to start growing and secrete estrogens (Figure 10-8). As the estrogen content of blood increases, it stimulates the anterior pituitary gland to secrete another hormone, namely LH (luteinizing hormone). LH causes maturing of a follicle and its ovum, ovulation (rupturing of mature follicle with ejection of ovum), and luteinization (formation of a golden body, the corpus luteum, in the ruptured follicle).

Which hormone—FSH or LH—would you call the "ovulating hormone"? Do you think ovulation could occur if the blood concentration of FSH remained low throughout

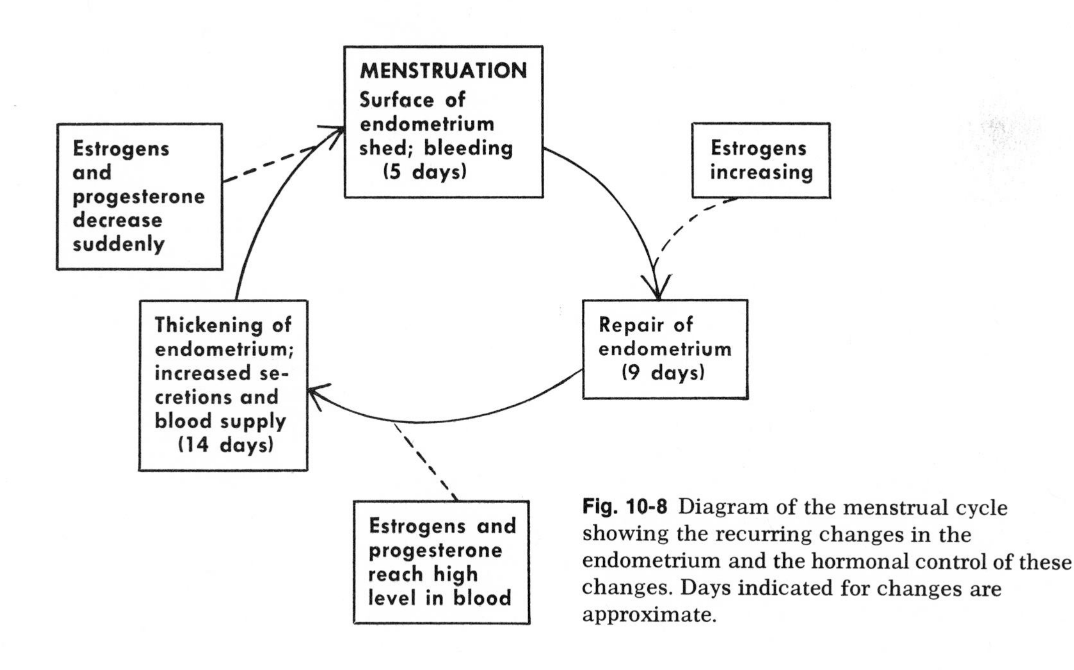

Fig. 10-8 Diagram of the menstrual cycle showing the recurring changes in the endometrium and the hormonal control of these changes. Days indicated for changes are approximate.

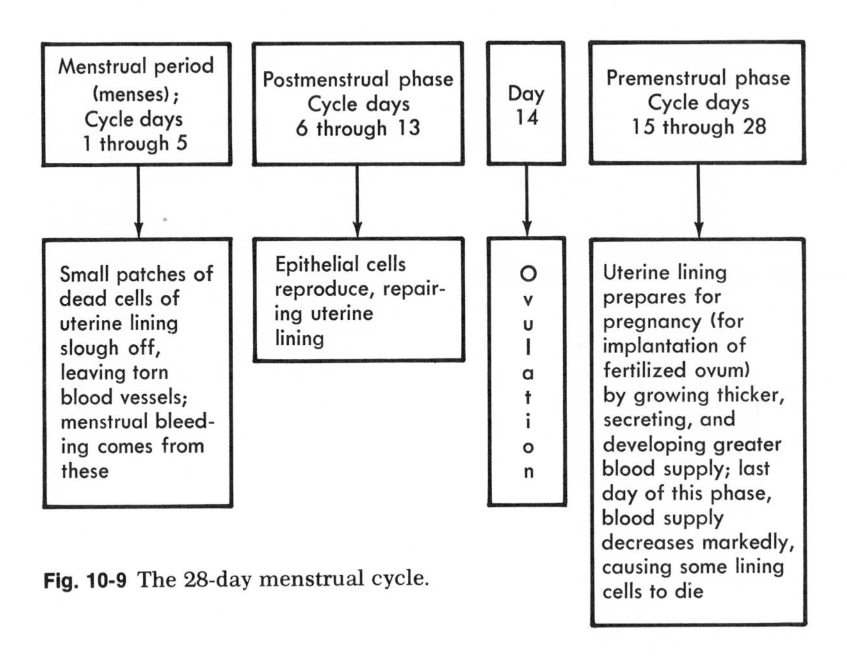

Fig. 10-9 The 28-day menstrual cycle.

the menstrual cycle? Ovulation cannot occur if the blood level of FSH stays low, because a high concentration of this hormone is essential to stimulate graafian follicles to start growing and maturing. With a low level of FSH no follicles start to grow, and therefore none become ripe enough to ovulate. Ovulation is caused by the combined actions of both FSH and LH. Birth control pills that contain estrogen substances suppress FSH secretion. This indirectly prevents ovulation.

As a general rule, only one ovum matures each month during the 30 or 40 years that a woman has menstrual periods. But there are exceptions to this rule. Some months more than one matures, and some months no ovum matures. Ovulation occurs about 14 days before the next menstrual period begins. In a 28-day menstrual cycle, this means that ovulation occurs on the fourteenth day of the cycle, as shown in Figure 10-9. (Note that the first day of the menstrual period is considered the first day of the cycle.) In a 30-day cycle, however, the fourteenth day before the beginning of the next menses is not the fourteenth cycle day, but the sixteenth. And in a 25-day cycle, the eleventh cycle day is the fourteenth day before the next menses begins.

This matter of the time of ovulation has great practical importance. An ovum lives only a short time after it is ejected from its follicle, and sperm live only a short time after they enter the female body. Fertilization of an ovum by a sperm, therefore, can occur only around the time of ovulation. In other words, a woman's fertile period lasts only a few days out of each month. This knowledge forms the basis of the rhythm method of birth control. Recently someone suggested calling this the "natural family planning" method.

Ovulation occurs, as we have said, because of the combined actions of the two anterior pituitary hormones, FSH and LH. The next question is what causes menstruation? A brief answer is this: a sudden, sharp decrease in estrogen and progesterone secretion toward the end of the premenstrual period causes the uterine lining to break down and another menstrual period to begin. But exactly what causes the sudden decrease in estrogen and progesterone secretion, we do not yet know.

Disorders of the reproductive systems

Infections of the mucous membrane lining of the reproductive tract occur in both sexes. In women the danger of this becomes especially great following childbirth. Physicians and nurses therefore take great care not to introduce infectious organisms into the reproductive tract during or after delivery. The name given the inflammation indicates the part of the tract inflamed. For example, vaginitis is inflammation of the vagina; cervicitis inflammation of the cervix; endometritis, inflammation of the endometrium; salpingitis, inflammation of the uterine tubes. Tumors frequently develop in the uterus, ovaries, and breasts of women, and in the prostate gland of men.

outline summary

MALE REPRODUCTIVE SYSTEM

Structural plan

Consists of pair of main sex glands, series of ducts, accessory sex glands, and external reproductive organs (genitals); see Table 10-1

Structures

1 External genitals – the scrotum and penis
2 Testes – pair of small oval glands in scrotum; main male sex glands; seminiferous tubules form male sex cells (spermatogenesis); interstitial cells secrete male hormone testosterone
3 Epididymis – narrow tube attached to each testis; duct of testis
4 Vas deferens – continuation of duct that starts with epididymis
5 Ejaculatory duct – continuation of seminal duct
6 Urethra – terminal duct of male reproductive tract as well as of urinary tract
7 Seminal vesicles – accessory gland; secrete alkaline substance into semen
8 Prostate gland – encircles urethra just below bladder; secretes alkaline fluid, part of semen
9 Bulbourethral glands – pair of small glands located just below prostate; ducts open into urethra where they add alkaline secretion to semen

FEMALE REPRODUCTIVE SYSTEM

Structural plan

Same as male; see Table 10-2

Structures

1 External genitals, vulva – see Figure 10-3
2 Ovaries – main sex glands of female; located in pelvic cavity; female sex cells (ova) form in graafian follicles in ovaries; graafian follicles secrete estrogens (see Figure 10-4 for estrogen functions); corpus luteum secretes estrogens and progesterone (see Figure 10-5 for functions)
3 Uterine tubes (fallopian tubes) – ducts for ovaries but not attached to them
4 Uterus (womb)
   a Parts – fundus, body, cervix; see Figure 10-6
   b Structure – strong muscular organ with mucous lining (endometrium)
   c Functions – menstruation, pregnancy, labor
5 Vagina – muscular tube lined with mucous membrane; terminal part of female reproductive tract
6 Bartholin's glands – pair of small glands near vaginal orifice; secrete lubricating fluid
7 Breasts – glands present in both sexes but normally function only in female

Menstrual cycle

1 Length – about 28 days, varies somewhat in different individuals and in the same individual at different times
2 Phases
   a Menstrual period or menses – about the first 4 or 5 days of the cycle, varies somewhat; characterized by sloughing of bits of endometrium (uterine lining) with bleeding
   b Postmenstrual phase – days between the end of menses and ovulation; varies in length; the shorter the cycle, the shorter the postmenstrual phase; the longer the cycle, the longer the postmenstrual phase; examples, in 28-day cycle postmenstrual phase ends on thirteenth day, but in 26-day cycle it ends on eleventh day, and in 32-day cycle, it ends on seventeenth day; characterized by repair of endometrium
   c Premenstrual phase – days between ovulation and beginning of next menses; ovulation about 14 days before next menses; characterized by

further thickening of endometrium and secretion by its glands in preparation for implantation of fertilized ovum; combined actions of the anterior pituitary hormones FSH and LH cause ovulation; sudden, sharp decrease in estrogens and progesterone bring on menstruation if pregnancy does not occur

DISORDERS OF THE REPRODUCTIVE SYSTEMS

1 Infections—common in both sexes; for example, salpingitis, vaginitis, endometritis, prostatitis
2 Tumors—occur in both sexes; for example, in uterus, ovaries, breasts, and prostate

new words

cervicitis
climacteric
endometritis
menarche
menopause
menses
ovulation
prostatitis
vaginitis

review questions

1 Identify the masculinizing hormone by its scientific name. What glands secrete it?
2 Identify the feminizing hormone by its scientific name. What gland secretes it?
3 Identify the ovulating hormone by its scientific name. What glands secrete it?
4 Identify and locate each of the following:

Bartholin's glands	ovary
cervix	ovum
epididymis	sperm
graafian follicles	testes

5 Define briefly the words listed under "new words."
6 What causes ovulation?
7 What causes menstruation?
8 How many female sex cells are usually formed each month? How does this compare with the number of male sex cells formed each month?

Suggested supplementary readings

Chapter 1

Anthony, C. P.: Basic concepts in anatomy and physiology—a programmed presentation, ed. 2, St. Louis, 1970, The C. V. Mosby Co., pp. 1-7.

Anthony, C. P., and Kolthoff, N. J.: Textbook of anatomy and physiology, ed. 8, St. Louis, 1971, The C. V. Mosby Co., pp. 2-60.

Francis, C. C: Introduction to human anatomy, ed. 5, St. Louis, 1968, The C. V. Mosby Co., pp. 22-36, 388-399.

Montagna, W.: The skin, Sci. Amer. 212:56-66, 1965.

Chapter 2

Anthony, C. P., and Kolthoff, N. J.: Textbook of anatomy and physiology, ed. 8, St. Louis, 1971, The C. V. Mosby Co., pp. 63-111.

Francis, C. C: Introduction to human anatomy, ed. 5, St. Louis, 1968, The C. V. Mosby Co., pp. 39-125.

Loomis, W. F.: Rickets, Sci. Amer. 223:77-91, 1970.

Chapter 3

Anthony, C. P., and Kolthoff, N. J.: Textbook of anatomy and physiology, ed. 8, St. Louis, 1971, The C. V. Mosby Co., pp. 113-160.

Francis, C. C: Introduction to human anatomy, ed. 5, St. Louis, 1968, The C. V. Mosby Co., pp. 126-178.

Olson, E. V., and Edmonds, R. E.: The hazards of immobility, effects on motor function, Amer. J. Nurs. 67(4):788-790, 1967.

Chapter 4

Anthony, C. P.: Basic concepts in anatomy and physiology—a programmed presentation, ed. 2, St. Louis, 1970, The C. V. Mosby Co., pp. 51-90.

Anthony, C. P., and Kolthoff, N. J.: Textbook of anatomy and physiology, ed. 8, St. Louis, 1971, The C. V. Mosby Co., pp. 164-262.

Dean, G.: The multiple sclerosis problem, Sci. Amer. **223**(1):40-46, 1970.
Plummer, E. M.: The MS patient, Amer. J. Nurs. **68**(10): 2161-2167, 1968.
Haynes, U. H.: Nursing approaches in cerebral dysfunction, Amer. J. Nurs. **68**(10):2170-2176, 1968.
Ratcliff, J. D.: I am Joe's hypothalamus, Reader's Digest, March, 1970, pp. 124-127.

Chapter 5

Anthony, C. P.: Basic concepts in anatomy and physiology—a programmed presentation, ed. 2, St. Louis, 1970, The C. V. Mosby Co., pp. 91-107.
Anthony, C. P., and Kolthoff, N. J.: Textbook of anatomy and physiology, ed. 8, St. Louis, 1971, The C. V. Mosby Co., pp. 263-289.
Rasmussen, H., and Pechet, M.: Calcitonin, Sci. Amer. **223**(4):42-50, 1970.

Chapter 6

Anthony, C. P.: Basic concepts in anatomy and physiology—a programmed presentation, ed. 2, St. Louis, 1970, The C. V. Mosby Co., pp. 22-50.
Anthony, C. P., and Kolthoff, N. J.: Textbook of anatomy and physiology, ed. 8, St. Louis, 1971, The C. V. Mosby Co., pp. 389-444.
Francis, C. C: Introduction to human anatomy, ed. 5, St. Louis, 1968, The C. V. Mosby Co., pp. 338-375.
Hayter, J.: Impaired liver function and related nursing care, Amer. J. Nurs. **68**(11):2374-2379, 1968.
Stokes, S. A.: Fasting for obesity, Amer. J. Nurs. **69**(4): 796-799, 1969.

Chapter 7

Anthony, C. P., and Kolthoff, N. J.: Textbook of anatomy and physiology, ed. 8, St. Louis, 1971, The C. V. Mosby Co., pp. 362-388.
Comroe, J. H., Jr.: The lung, Sci. Amer. **214**:57-66, 1966.
Francis, C. C: Introduction to human anatomy, ed. 5, St. Louis, 1968, The C. V. Mosby Co., pp. 312-337.
Nett, L. M., and Petty, T. L.: Why emphysema patients are the way they are, Amer. J. Nurs. **70**(6):1251-1255, 1970.
Winter, P. M., and Lowenstein, E.: Acute respiratory failure, Sci. Amer. **221**:23-29, 1969.

Chapter 8

Adolph, E. F.: The heart's pacemakers, Sci. Amer. **213**: 32-37, 1967.
Anthony, C. P.: Basic concepts in anatomy and physiology—a programmed presentation, ed. 2, St. Louis, 1970, The C. V. Mosby Co., pp. 108-144.
Anthony, C. P., and Kolthoff, N. J.: Textbook of anatomy and physiology, ed. 8, St. Louis, 1971, The C. V. Mosby Co., pp. 292-361.
Effler, D. B.: Surgery for coronary disease, Sci. Amer. **219**:36-43, 1968.
Francis, C. C: Introduction to human anatomy, ed. 5, St. Louis, 1968, The C. V. Mosby Co., pp. 261-311.

Chapter 9

Anthony, C. P.: Basic concepts in anatomy and physiology—a programmed presentation, ed. 2, St. Louis, 1970, The C. V. Mosby Co., pp. 145-149.
Anthony, C. P., and Kolthoff, N. J.: Textbook of anatomy and physiology, ed. 8, St. Louis, 1971, The C. V. Mosby Co., pp. 446-463.
Cummings, J. W.: Hemodialysis—feelings, facts, and fantasies, Amer. J. Nurs. **70**(1):70-83, 1970.
Francis, C. C: Introduction to human anatomy, ed. 5, St. Louis, 1968, The C. V. Mosby Co., pp. 376-387.
Ratcliff, J. D.: I am Joe's kidney, Reader's Digest, May, 1970, pp. 98-102.

Chapter 10

Anthony, C. P., and Kolthoff, N. J.: Textbook of anatomy and physiology, ed. 8, St. Louis, 1971, The C. V. Mosby Co., pp. 466-494.
Connell, E. B.: The pill and the problems, Amer. J. Nurs. **71**(2):326-332, 1971.
Francis, C. C: Introduction to human anatomy, ed. 5, St. Louis, 1968, The C. V. Mosby Co., pp. 405-421.
Gonzales, B.: Voluntary sterilization, Amer. J. Nurs. **70**(12): 2581-2583, 1970.

Glossary

abdomen (ab-do'men) body area between the diaphragm and pelvis.

abduct (ab-dukt') to move away from the midline; opposite of adduct.

absorption (ab-sorp'shun) passage of a substance through a membrane (for example, skin or mucosa) into blood.

acetabulum (as"e-tab'u-lum) socket in the hip bone (os coxa or innominate bone) into which the head of the femur fits.

Achilles tendon (ah-kil'ēz ten'dun) tendon inserted on calcaneus; so-called because of the Greek myth that Achilles' mother held him by the heels when she dipped him in the river Styx, thereby making him invulnerable except in this area.

acidosis (as"i-do'sis) condition in which there is an excessive proportion of acid in the blood.

acromion (ah-kro'me-on) bony projection of the scapula; forms point of the shoulder.

adduct (ah-dukt') to move toward the midline; opposite of abduct.

adenohypophysis (ad"ĕ-no-hi-pof'i-sis) anterior pituitary gland.

adenoid (ad'ĕ-noid) literally, glandlike; adenoids, or pharyngeal tonsils, are paired lymphoid structures in the nasopharynx.

adolescence (ad"o-les'ens) period between puberty and adulthood.

adrenergic fibers (ad"ren-er'jik fi'bers) axons whose terminals release norepinephrine and epinephrine.

afferent neuron (af'er-ent nu'ron) transmitting impulses to the central nervous system.

albuminuria (al"bumi-nu-re-ah) albumin in the urine.

alkalosis (al"kah-lo'sis) condition in which there is an excessive proportion of alkali in the blood; opposite of acidosis.

alveolus (al-ve'o-les) literally a small cavity; alveoli of lungs are microscopic saclike dilatations of terminal bronchioles.

amenorrhea (ah-men″o-re′ah) absence of the menses.

amino acid (am′ĭ-no as′id) organic compound having an NH_3 and a COOH group in its molecule; has both acid and basic properties; amino acids are the structural units from which proteins are built.

amphiarthrosis (am″fe-ar-thro′sis) slightly movable joint.

anabolism (ah-nab′o-lizm) synthesis by cells of complex compounds (for example, protoplasm and hormones) from simpler compounds (amino acids, simple sugars, fats, and minerals); opposite of catabolism, the other phase of metabolism.

anastomosis (ah-nas″to-mo′sis) connection between vessels; the circle of Willis, for example, is an anastomosis of certain cerebral arteries.

anemia (ah-ne′me-ah) deficient number of red blood cells or deficient hemoglobin.

anesthesia (an″es-the′ze-ah) loss of sensation.

aneurysm (an′u-rizm) blood-filled saclike dilatation of the wall of an artery.

angina (an′jĭ-nah) any disease characterized by spasmodic suffocative attacks; for example, angina pectoris and paroxysmal thoracic pain with feeling of suffocation.

ankylosis (ang″kĭ-lo′sis) abnormal immobility of a joint.

anorexia (an″o-rek′se-ah) loss of appetite.

anoxia (an-ok′se-ah) deficient oxygen supply to tissues.

antagonistic muscles (an-tag′o-nis-tik mus′el) those having opposing actions; for example, muscles that flex the upper arm are antagonists to muscles that extend it.

anterior (an-te′re-or) front or ventral; opposite of posterior or dorsal.

antibody, immune body (an′tĭ-bod″e, ĭ-mūn′ bod′e) substance produced by the body that destroys or inactivates a specific substance (antigen) that has entered the body; for example, diphtheria antitoxin is the antibody against diphtheria toxin.

antigen (an′tĭ-jen) substance that, when introduced into the body, causes formation of antibodies against it.

antiseptic (an″tĭ-sep′tik) preventing bacterial growth and multiplication.

antrum (an′trum) cavity; for example, the antrum of Highmore, the space in each maxillary bone, or the maxillary sinus.

anus (a′nus) distal end or outlet of the rectum.

apex (a′peks) pointed end of a conical structure.

aphasia (ah-fa′ze-ah) loss of a language faculty, such as the ability to use words or to understand them.

apnea (ap-ne′ah) temporary cessation of breathing.

aponeurosis (ap″o-nu-ro′sis) flat sheet of white fibrous tissue that serves as a muscle attachment.

arachnoid (ah-rak′noid) delicate, weblike middle membrane of the meninges.

areola (ah-re′o-lah) small space; the pigmented ring around the nipple.

arteriole (ar-te′re-ōl) small branch of an artery.

artery (ar′ter-e) vessel carrying blood away from the heart.

arthrosis (ar-thro′sis) joint or articulation.

articulation (ar-tik-u-la′shun) joint.

ascites (ah-si′tēz) accumulation of serous fluid in the abdominal cavity.

asphyxia (as-fik′se-ah) loss of consciousness caused by deficient oxygen supply.

aspirate (as′pŭ-rāt) to remove by suction.

ataxia (ah-tak′se-ah) loss of power of muscle coordination.

atrium (a′tre-um) chamber or cavity; for example, atrium of each side of the heart.

atrophy (at′ro-fe) wasting away of tissue; decrease in size of a part.

auricle (aw′re-kl) part of the ear attached to the side of the head; earlike appendage of each atrium of heart.

autonomic (aw″to-nom′ik) self-governing; independent.

axilla (ak-sil′ah) armpit.

axon (ak′son) nerve cell process that transmits impulses away from the cell body.

Bartholin (bar′to-lin) seventeenth century Danish anatomist.

basophil (ba′so-fil) white blood cell that stains readily with basic dyes.

biceps (bi′seps) two headed.

bilirubin (bil″e-roo′bin) red pigment in the bile.

biliverdin (bil″e-ver′din) green pigment in the bile.

Bowman (Bo′man) nineteenth century English physician.

brachial (bra′ke-al) pertaining to the arm.

bronchiectasis (brong″ke-ek′-tah-sis) dilatation of the bronchi.

bronchiole (brong′ke-ōl) small branch of a bronchus.

bronchus (brong′kus) one of the two branches of the trachea.

buccal (buk′al) pertaining to the cheek.

buffer (buf′er) compound that combines with an acid or with a base to form a weaker acid or base, thereby lessening the change in hydrogen-ion concentration that would occur without the buffer.

bursa (bur′sah) fluid-containing sac or pouch lined with synovial membrane.

buttock (but′ok) prominence over the gluteal muscles.

calculus (kal′ku-lus) stone; formed in various parts of the body; may consist of different substances.

calorie (kal′o-re) heat unit; a large calorie is the amount of heat needed to raise the temperature of 1 kilogram of water 1 degree centigrade.

calyx (ka′liks) cup-shaped division of the renal pelvis.

capillary (kap′i-lār″e) microscopic blood vessel; capillaries connect arterioles with venules; also, microscopic lymphatic vessels.

carbaminohemoglobin (kar-bam″ĭ-no-he″mo-glo′bin) compound formed by union of carbon dioxide with hemoglobin.

carbohydrate (kar″bo-hi′drāt) organic compounds containing carbon, hydrogen, and oxygen in certain specific proportions; for example, sugars, starches, and cellulose.

carboxyhemoglobin (kar-bok″se-he″mo-glo′-bin) compound formed by union of carbon monoxide with hemoglobin.

carcinoma (kar″sĭ-no′mah) cancer, a malignant tumor.

caries (ka′re-ēz) decay of teeth or of bone.

carotid (kah-rot′id) from Greek word meaning to plunge into deep sleep; carotid arteries of the neck so called because pressure on them may produce unconsciousness.

carpal (kar′pal) pertaining to the wrist.

casein (ka′se-in) protein in milk.

cast (kast) mold; for example, formed in renal tubules.

castration (kas-tra′shun) removal of testes or ovaries.

catabolism (kah-tab′o-lizm) breakdown of food compounds or of protoplasm into simpler compounds; opposite of anabolism, the other phase of metabolism.

catalyst (kat′ah-list) substance that accelerates the rate of a chemical reaction.

cataract (kat′ah-rakt) opacity of the lens of the eye.

catecholamines (kat″e-kol-am′inz) norepinephrine and epinephrine.

cecum (se′kum) blind pouch; the pouch at the proximal end of the large intestine.

celiac (se′le-ak) pertaining to the abdomen.

cellulose (sel′u-lōs) polysaccharide, the main plant carbohydrate.

centimeter (sen′tĭ-me″ter) 1/100 of a meter, about 2/5 of an inch.

centrioles (sen′trĭ-ōlz) two dots seen with a light microscope in the centrosphere of a cell; active during mitosis.

cerumen (sĕ-roo′men) earwax.

cervix (ser′viks) neck; any neckline structure.

chiasm (ki′azm) crossing; specifically, a crossing of the optic nerves.

cholecystectomy (ko″le-sis-tek′to-me) removal of the gallbladder.

cholesterol (ko-les′ter-ol) organic alcohol present in bile, blood, and various tissues.

cholinergic fibers (ko″lin-er′jik fi′bers) axons whose terminals release acetylcholine.

cholinesterase (ko″lin-es′ter-ās) enzyme; catalyzes breakdown of acetylcholine.

chromatin (kro′mah-tin) deep-staining substance in the nucleus of cells; divides into chromosomes during mitosis.

chromosome (kro′mo-sōm) one of the segments into which chromatin divides during mitosis; involved in transmitting hereditary characteristics.

chyle (kīl) milky fluid; the fat-containing lymph in the lymphatics of the intestine.

chyme (kīm) partially digested food mixture leaving the stomach.

cilia (sil′e-ah) hairlike projections of protoplasm.

circadian (ser″kah-de′an) daily.

cochlea (kok′le-ah) snail shell or structure of similar shape.

coenzyme (ko-en′zim) nonprotein substance that activates an enzyme.

collagen (kol′ah-jen) principle organic constituent of connective tissue.

colloid (kol′oid) dissolved particles with diameters of 1 to 100 millimicrons (1 millimicron equals about 1/25,000,000 of an inch).

colostrum (ko-los′trum) first milk secreted after childbirth.

concha (kong′kah) shell-shaped structure; for example, bony projections into the nasal cavity.

condyle (kon′dīl) rounded projection at the end of a bone.

congenital (kon-jen′ĭ-tal) present at birth.

contralateral (kon″trah-lat′er-al) on the opposite side.

coracoid (kor′ah-koid) like a raven's beak in form.

corium (ko′re-um) true skin or derma.

coronal (ko-ro′nal) like a crown.

coronary (kor′o-na-re) encircling; in the form of a crown.

corpus (kor′pus) body.

corpuscle (kor′pus′l) very small body or particle.

cortex (kor′teks) outer part of an internal organ; for example, of the cerebrum and of the kidneys.

cortisol (kor′ti-sol) the chief hormone secreted by the adrenal cortex; hydrocortisone; compound F.

costal (kos′tal) pertaining to the ribs.

crenation, plasmolysis (kre-na′shun, plaz-mol′ĭ-sis) shriveling of a cell caused by water withdrawal.

cretinism (kre′tin-izm) dwarfism caused by hypofunction of the thyroid gland.

cribriform (krib′rĭ-form) sievelike.

cricoid (kri′koid) ring-shaped; a cartilage of this shape in the larynx.

crystalloid (kris′tal-loid) dissolved particle less than 1 millimicron in diameter.

cutaneous (ku-ta′ne-us) pertaining to the skin.

cyanosis (si″ah-no′sis) bluish appearance of the skin caused by deficient oxygenation of the blood.

cytology (si-tol′o-je) study of cells.

cytoplasm (si′to-plazm″) the protoplasm of a cell exclusive of the nucleus.

deciduous (de-sid′u-us) temporary; shedding at a certain stage of growth; for example, deciduous teeth.

decussation (de″kus-sa′shun) crossing over like an X.

defecation (def″e-ka′shun) elimination of waste matter from the intestines.

deglutition (deg″loo-tish′un) swallowing.

deltoid (del′toid) triangular; for example, deltoid muscle.

dendrite, dendron (den′drīt, den′dron) branching or treelike; a nerve cell process that transmits impulses toward the cell body.

dens (dens) tooth.

dentate (den′tāt) having toothlike projections.

dentine (den′tēn) main part of a tooth, under the enamel.

dentition (den-tish′un) teething; also, number, shape, and arrangement of the teeth.

dermis, corium (der′mis, ko′re-um) true skin.

dextrose (deks′trōs) glucose, a monosaccharide, the principal blood sugar.

diaphragm (di′ah-fram) membrane or partition that separates one thing from another; the muscular partition between the thorax and abdomen; the midriff.

diaphysis (di-af′ĭ-sis) shaft of a long bone.

diarthrosis (di'ar-thro'sis) freely movable joint.
diastole (di-as'to-le) relaxation of the heart, interposed between its contractions; opposite of systole.
diencephalon (di″en-sef'ah-lon) "tween" brain; parts of the brain between the cerebral hemispheres and the mesencephalon or midbrain.
diffusion (dĭ-fu'zhun) spreading; for example, scattering of dissolved particles.
digestion (di-jest'yun) conversion of food into assimilable compounds.
diplopia (dĭ-plo'pe-ah) double vision; seeing one object as two.
disaccharide (di-sak'ah-rid') sugar formed by the union of two monosaccharides; contains twelve carbon atoms.
distal (dis'tal) toward the end of a structure; opposite of proximal.
dorsal, posterior (dor'sal, pos-te're-or) pertaining to the back; opposite of ventral.
dropsy (drop'se) accumulation of serous fluid in a body cavity or in tissues; edema.
dura mater (du'rah ma'ter) literally strong or hard mother; outermost layer of the meninges.
dyspnea (disp'ne-ah) difficult or labored breathing.
dystrophy (dis'tro-fe) faulty nutrition.

ectopic (ek-top'ik) displaced; not in the normal place; for example, extrauterine pregnancy.
edema (e-de'mah) excessive fluid in tissues; dropsy.
effector (ef-fek'tor) responding organ; for example, voluntary and involuntary muscle, the heart, and glands.
efferent (ef'er-ent) carrying from, as neurons that transmit impulses from the central nervous system to the periphery; opposite of afferent.
electrocardiogram (e-lek″tro-kar'de-o-gram) graphic record of heart's action potentials.
electroencephalogram (e-lek″tro-en-sef'ah-lo-gram) graphic record of heart's action potentials.
electrolyte (e-lek'tro-līt) substance that ionizes in solution, rendering the solution capable of conducting an electric current.
electron (e-lek'tron) minute, negatively charged particle.
elimination (e-lim″ĭ-na'shun) expulsion of wastes from the body.
embolism (em'bo-lizm) obstruction of a blood vessel by foreign matter carried in the bloodstream.
embryo (em'bre-o) animal in early stages of intrauterine development; the human fetus the first three months after conception.
emesis (em'e-sis) vomiting.
emphysema (em″fi-se'mah) dilatation of pulmonary alveoli.
empyema (em″pi-e'mah) pus in a cavity; for example, in the chest cavity.
encephalon (en-sef'ah-lon) brain.
endocrine (en'do-krīn) secreting into the blood or tissue fluid rather than into a duct; opposite of exocrine.
energy (en'er-je) capacity for doing work.
enteron (en'ter-on) intestine.
enzyme (en'zīm) catalytic agent formed in living cells.
eosinophil, acidophil (e″o-sin'o-fil, ah-sid'o-fil) white blood cell readily stained by eosin.
epidermis (ep″ĭ-der'mis) "false" skin; outermost layer of the skin.
epinephrine (ep″ı-nef'rin) adrenaline; secretion of the adrenal medulla.
epiphyses (e-pif'ĭ-sēz) ends of a long bone.
erythrocyte (e-rith'ro-sīt) red blood cell.
ethmoid (eth'moid) sievelike.
eupnea (ūp-ne'ah) normal respiration.
Eustachio (u-sta'she-o) Italian anatomist of the sixteenth century.
exocrine (ek-so'krin) secreting into a duct; opposite of endocrine.
exophthalmos (ek″sof-thal'mos) abnormal protrusion of the eyes.
extrinsic (eks-trin'sik) coming from the outside; opposite of intrinsic.

Fallopius (fal-lo'pe-us) sixteenth century Italian anatomist.
fascia (fash'e-ah) sheet of connective tissue.
fasciculus (fah-sik'u-lus) little bundle.
fetus (fe'tus) unborn young, especially in the later stages; in human beings, from third month of intrauterine period until birth.
fiber (fi'ber) threadlike structure.
fibrin (fi'brin) insoluble protein in clotted blood.
fibrinogen (fi-brin'o-jen) soluble blood protein that is converted to insoluble fibrin during clotting.
fimbria (fim'bre-ah) fringe.
fissure (fish'ūr) groove.
flaccid (flak'sid) soft, limp.
follicle (fol'lĭ-k'l) small sac or gland.
fontanelles (fon″tah-nelz') "soft spots" of the infant's head; unossified areas in the infant skull.
foramen (fo-ra'men) small opening.
fossa (fos'sah) cavity or hollow.
fovea (fo've-ah) small pit or depression.
fundus (fun'dus) base of a hollow organ; for example, the part farthest from its outlet.

ganglion (gang'gle-on) cluster of nerve cell bodies outside the central nervous system.
gasserian (gas-se're-an) named for Gasser, a sixteenth century Austrian surgeon.
gastric (gas'trik) pertaining to the stomach.
gene (jēn) part of the chromosome that transmits a given hereditary trait.
genitals (jen'i-tals) reproductive organs; genitalia.
gestation (jes-ta'shun) pregnancy.
gland (gland) secreting structure.
glomerulus (glo-mer'u-lus) compact cluster; for example, of capillaries in the kidneys.
glossal (glos'al) of the tongue.
glucagon (gloo'kah-gon) hormone secreted by alpha cells of the islands of Langerhans.
glucocorticoids (gloo″ko-kor'tĭ-koids) hormones that influence food metabolism; secreted by adrenal cortex.
gluconeogenesis (gloo″ko-ne″o-jen'e-sis) formation of glucose from protein or fat compounds.

glucose (gloo'kōs) monosaccharide or simple sugar; the principal blood sugar.
gluteal (gloo'te-al) of or near the buttocks.
glycerin, glycerol (glis'er-in, glis'er-ol) product of fat digestion.
glycogen (gli'ko-jen) polysaccharide; animal starch.
glycogenesis (gli"ko-jen'e-sis) formation of glycogen from glucose or from other monosaccharides, fructose or galactose.
glycogenolysis (gli"ko-jĕ-nol'ĭ-sis) hydrolysis of glycogen to glucose-6-phosphate or to glucose.
glyconeogenesis (gli"ko-ne"o-jen'e-sis) the formation of glycogen from protein or fat compounds.
gonad (gon'ad) sex gland in which reproductive cells are formed.
graafian (graf'e-an) named for Graaf, a seventeenth century Dutch anatomist.
gradient (gra'de-ent) a slope or difference between two levels; for example, blood pressure gradient—a difference between the blood pressure in two different vessels.
gustatory (gus'tah-to"re) pertaining to taste.
gyrus (ji'rus) convoluted ridge.

haversian (ha-ver'shan) named for Havers, English anatomist of the late seventeenth century.
hemiplegia (hem"e-ple'je-ah) paralysis of one side of the body.
hemoglobin (he"mo-glo'bin) iron-containing protein in red blood cells.
hemolysis (he-mol'ĭ-sis) destruction of red blood cells with escape of hemoglobin from them into surrounding medium.
hemopoiesis (he"mo-poi-e'sis) blood cell formation.
hemorrhage (hem'or-ij) bleeding.
hepar (he'par) liver.
heparin (hep'ah-rin) substances obtained from the liver; inhibits blood clotting.
heredity (he-red'ĭ-te) transmission of characteristics from a parent to a child.
hernia, "rupture" (her'ne-ah, rup'chur) protrusion of a loop of an organ through an abnormal opening.
hilus, hilum (hi'lus, hi'lum) depression where vessels enter an organ.
His (his) German anatomist of the late nineteenth century.
histology (his-tol'o-je) science of minute structure of tissues.
homeostasis (ho"me-o-sta'sis) relative uniformity of the normal body's internal environment.
hormone (hor'mōn) substance secreted by an endocrine gland.
hyaline (hi'ah-līn) glasslike.
hydrocortisone (hī"dro-kor'tĭ-sōn) a hormone secreted by the adrenal cortex; cortisol; compound F.
hydrolysis (hi-drol'ı-sis) literally "split by water"; chemical reaction in which a compound reacts with water to form simpler compounds.
hymen (hi'men) Greek for skin; mucous membrane that may partially or entirely occlude the vaginal outlet.
hyoid (hi'oid) U-shaped; bone of this shape at the base of the tongue.
hyperemia (hi"per-e'me-ah) increased blood in a part of the body.
hyperkalemia (hi"per-kah-le'me-ah) higher than normal concentration of potassium in the blood.
hypernatremia (hi"per-nah-tre'me-ah) higher than normal concentration of sodium in the blood.
hyperopia (hi"per-o'pe-ah) farsightedness.
hyperplasia (hi"per-pla'ze-ah) increase in the size of a part caused by an increase in the number of its cells.
hyperpnea (hi"perp-ne'ah) abnormally rapid breathing; panting.
hypertension (hi"per-ten'shun) abnormally high blood pressure.
hyperthermia (hi"per-ther'me-ah) fever; body temperature above 37° C.
hypertrophy (hi-per'tro-fe) increased size of a part caused by an increase in the size of its cells.
hypokalemia (hi"po-ka-le'me-ah) lower than normal concentration of potassium in the blood.
hyponatremia (hi"po-nah-tre'me-ah) lower than normal concentration of sodium in the blood.
hypophysis (hi-pof'ĭ-sis) Greek for undergrowth; hence the pituitary gland, which grows out from the undersurface of the brain.
hypothermia (hi"po-ther'me-ah) subnormal body temperature below 37° C.
hypoxia (hi-pok'se-ah) oxygen deficiency.

incus (ing'kus) anvil; the middle ear bone that is shaped like an anvil.
inferior (in-fe're-or) lower; opposite of superior.
inguinal (ing'gwĭ-nal) of the groin.
inhalation (in"hah-la'shun) inspiration or breathing in; opposite of exhalation or expiration.
inhibition (in"hi-bish'un) checking or restraining of action.
innominate (ĭ-nom'ĭ-nāt) not named, anonymous; for example, ossa coxae (hip bones) formerly known as innominate bones.
insulin (in'su-lin) hormone secreted by islands of Langerhans in the pancreas.
intercellular (in"ter-sel'u-lar) between cells; interstitial.
internuncial (in"ter-nun'she-al) like a messenger between two parties; hence an internuncial neuron (or interneuron) is one that conducts impulses from one neuron to another.
interstitial (in"ter-stish'al) forming small spaces between things; intercellular.
intrinsic (in-trin'sik) not dependent upon externals; located within something; opposite of extrinsic.
involuntary (in-vol'un-ter"e) not willed; opposite of voluntary.
involution (in"vo-lu'shun) return of an organ to its normal size after enlargement; also retrograde or degenerative change.
ion (i'on) electrically charged atom or group of atoms.
ipsilateral (ip"sĭ-lat'er-al) on the same side; opposite of contralateral.
irritability (ir"ĭ-tah-bil'ĭ-te) excitability; ability to react to a stimulus.
ischemia (is-ke'me-ah) local anemia; temporary lack of blood supply to an area.

isotonic (i″so-ton′ik) of the same tension or pressure.

ketones (ke′tōns) acids (acetoacetic, beta-hydroxybutyric, and acetone) produced during fat catabolism.
ketosis (ke-to′sis) excess amount of ketone bodies in the blood.
kilogram (kil′o-gram) 1,000 grams; approximately 2.2 pounds.
kinesthesia (kin″es-the′ze-ah) "muscle sense"; that is, sense of position and movement of body parts.

labia (la′be-ah) lips.
lacrimal (lak′ri-mal) pertaining to tears.
lactation (lak-ta′shun) secretion of milk.
lactose (lak′tōs) milk sugar, a disaccharide.
lacuna (lah-ku′nah) space or cavity; for example, lacunae in bone contain bone cells.
lamella (lah-mel′ah) thin layer, as of bone.
lateral (lat′er-al) of or toward the side; opposite of medial.
leukocyte (lu′ko-sīt) white blood cell.
ligament (lig′ah-ment) bond or band connecting two objects; in anatomy a band of white fibrous tissue connecting bones.
lipid (lip′id) fats and fatlike compounds.
loin (loin) part of the back between the ribs and hip bones.
lumbar (lum′ber) of or near the loins.
lumen (lu′men) passageway or space within a tubular structure.
luteum (lu′te-um) golden yellow.
lymph (limf) watery fluid in the lymphatic vessels.
lymphocyte (lim′fo-sīt) one type of white blood cells.
lysosomes (li′so-sōms) membranous organelles containing various enzymes that can dissolve most cellular compounds; hence called "digestive bags" or "suicide bags" of cells.

malleolus (mal-le′o-lus) small hammer; projections at the distal ends of the tibia and fibula.
malleus (mal′e-us) hammer; the tiny middle ear bone that is shaped like a hammer.
Malpighi (mal-pig′e) seventeenth century Italian anatomist.
maltose (mawl′tōs) disaccharide or "double" sugar.
mammary (mam′er-e) pertaining to the breast.
manometer (mah-nom′e-ter) instrument used for measuring the pressure of fluids.
manubrium (mah-nu′bre-um) handle; upper part of the sternum.
mastication (mas″ti-ka′shun) chewing.
matrix (ma′triks) ground substance in which cells are embedded.
meatus (me-a′tus) passageway.
medial (me′de-al) of or toward the middle; opposite of lateral.
mediastinum (me″de-as-ti′num) middle section of the thorax; that is, between the two lungs.
medulla (me-dul′lah) Latin for marrow; hence the inner portion of an organ in contrast to the outer portion, or cortex.
meiosis (mi-o′sis) nuclear division in which the number of chromosomes are reduced to half their original number before the cell divides in two.
membrane (mem′brān) thin layer or sheet.
menstruation (men″stroo-a′shun) monthly discharge of blood from the uterus.
mesentery (mes′en-ter″e) fold of peritoneum that attaches the intestine to the posterior abdominal wall.
mesial (me′ze-al) situated in the middle; median.
metabolism (mĕ-tab′o-lizm) complex process by which food is utilized by a living organism.
metacarpus (met″ah-kar′pus) "after" the wrist; hence the part of the hand between the wrist and fingers.
metatarsus (met″ah-tar′sus) "after" the instep; hence the part of the foot between the tarsal bones and toes.
meter (me′ter) about 39.5 inches.
microglia (mi-krog′le-ah) one type of connective tissue found in the brain and cord.
micron (mi′kron) 1/1,000 millimeter; about 1/25,000 inch.
micturition (mik″tu-rish′un) urination, voiding.
millimeter (mil′lĭ-me-ter) 1/1,000 meter; about 1/25 inch.
mineralocorticoids (min″er-al-o-kor′tĭ-koidz) hormones that influence mineral salt metabolism; secreted by adrenal cortex; aldosterone is the chief mineralocorticoid.
mitochondria (mi″to-kon′dre-ah) threadlike structures.
mitosis (mi-to′sis) indirect cell division involving complex changes in the nucleus.
mitral (mi′tral) shaped like a miter.
monosaccharide (mon″o-sak′ah-rīd) simple sugar; for example, glucose.
motoneurons (mo″to-nu′rons) transmit nerve impulses away from the brain or spinal cord; also called motor, or efferent, neurons.
myelin (mi′ĕ-lin) lipoid substance found in the myelin sheath around some nerve fibers.
myocardium (mi″o-kar′de-um) muscle of the heart.
myopia (mi-o′pe-ah) nearsightedness.

nares (na′rēz) nostrils.
neurilemma (nu″rī-lem′mah) nerve sheath.
neurohypophysis (nu″ro-hi-pof′ĭ-sis) posterior pituitary gland.
neuron (nu′ron) nerve cell, including its processes.
neutrophil (nu′tro-fil) white blood cell that stains readily with neutral dyes.
nucleus (nu′kle-us) spherical structure within a cell; a group of neuron cell bodies in the brain or cord.

occiput (ok′sĭ-put) back of the head.
olecranon (o-lek′rah-non) elbow.
olfactory (ol-fak′to-re) pertaining to the sense of smell.
ophthalmic (of-thal′mik) pertaining to the eyes.
organelle (or″gan-el′) cell organ; one of the specialized parts of a single-celled organism (protozoon), serving for the performance of some individual function.
os (os) Latin for mouth and for bone.
osmosis (os-mo′sis) movement of a fluid through a semipermeable membrane.
ossicle (os′sı-k′l) little bone.

oxidation (ok″sĭ-da′shun) loss of hydrogen or electrons from a compound or element.
oxyhemoglobin (ok″se-he″mo-glo′bin) a compound formed by union of oxygen with hemoglobin.

palate (pal′at) roof of the mouth.
palpebrae (pal′pe-bre) eyelids.
papilla (pah-pil′lah) small nipple-shaped elevation.
paralysis (pah-ral′ĭ-sis) loss of the power of motion or sensation, especially loss of voluntary motion.
parenchyma (par-eng′kĭ-mah) essential or functional tissue of an organ.
parietal (pah-ri′ĕ-tal) of the walls of an organ or cavity.
parotid (pah-rot′id) located near the ear.
parturition (par″tu-rish′un) act of giving birth to an infant.
patella (pah-tel′lah) small, shallow pan; the kneecap.
pectoral (pek′to-ral) pertaining to the chest or breast.
pelvis (pel′vis) basin or funnel-shaped structure.
peripheral (pĕ-rif′er-al) pertaining to an outside surface.
pH (pe′ach′) hydrogen-ion concentration.
phagocytosis (fag″o-si-to′sis) ingestion and digestion of particles by a cell.
phalanges (fe-lan′jēz) finger or toe bones.
phrenic (fren′ik) pertaining to the diaphragm.
pia mater (pi′ah ma′ter) gentle mother; the vascular innermost covering (meninges) of the brain and cord.
pineal (pin′e-al) shaped like a pine cone.
piriformis (pir″i-for′mis) pear-shaped.
pisiform (pi′sĭ-form) pea-shaped.
plantar (plan′tar) pertaining to the sole of the foot.
plasma (plaz′mah) liquid part of the blood.
plasmolysis (plaz-mol′ı-sis) shrinking of a cell caused by water loss by osmosis.
plexus (plek′sus) network.
polymorphonuclear (pol″e-mor″fo-nu′kle-ar) having many-shaped nuclei.
polysaccharide (pol″e-sak′ah-rīd) complex sugar.
pons (ponz) bridge.
popliteal (pop-lit′e-al) behind the knee.
posterior (pos-te′re-or) following after; hence, located behind; opposite of anterior.
presbyopia (pres″be-o′pe-ah) "oldsightedness"; farsightedness of old age.
pronate (pro′nāt) to turn palm downward.
proprioceptors (pro′pri-o-sep′tors) receptors located in the muscles, tendons, and joints.
protoplasm (pro′to-plazm) living substance.
proximal (prok′sĭ-mal) next or nearest; located nearest the center of the body or the point of attachment of a structure.
psoas (so′as) pertaining to the loin, the part of the back between the ribs and hip bones.
psychosomatic (si″ko-so-mat′ik) pertaining to the influence of the mind, notably the emotions, on body functions.
puberty (pu′ber-te) age at which the reproductive organs become functional.

receptor (ri-sep′tor) peripheral beginning of a sensory neuron's dendrite.
reflex (re′fleks) involuntary action.
refraction (re-frak′shun) bending of a ray of light as it passes from a medium of one density to one of a different density.
renal (re′nal) pertaining to the kidney.
reticular (re-tik′u-lar) netlike.
reticulum (re-tik′u-lum) a network.
ribosomes (ri′bo-sōms) organelles in cytoplasm of cells; synthesize proteins, so nicknamed "protein factories."
rugae (roo′jee) wrinkles or folds.

sagittal (saj′ı-tal) like an arrow; longitudinal.
salpinx (sal′pinks) tube; oviduct.
sartorius (sar-to′re-us) tailor; hence, the thigh muscle used to sit cross-legged like a tailor.
sciatic (si-at′ik) pertaining to the ischium.
sclera (skle′rah) from Greek for hard.
scrotum (skro′tum) bag.
sebum (se′bum) Latin for tallow; secretion of sebaceous glands.
sella turcica (sel′ah tur′sikah) Turkish saddle; saddle-shaped depression in the sphenoid bone.
semen (se′men) Latin for seed; male reproductive fluid.
semilunar (sem″e-lu′nar) half-moon shaped.
senescence (se-nes′ens) old age.
serratus (ser-ra′tus) saw-toothed.
serum (se′rum) any watery animal fluid; clear, yellowish liquid that separates from a clot of blood.
sesamoid (ses′ah-moid) shaped like a sesame seed.
sigmoid (sig′moid) S-shaped.
sinus (si′nus) cavity.
soleus (so′le-us) pertaining to a sole; a muscle in the leg shaped like the sole of a shoe.
somatic (so-mat′ik) of the body framework or walls, as distinguished from the viscera or internal organs.
sphenoid (sfe′noid) wedge-shaped.
sphincter (sfingk′ter) ring-shaped muscle.
splanchnic (splank′nik) visceral.
squamous (skwa′mus) scalelike.
stapes (sta′pēz) stirrup; tiny stirrup-shaped bone in the middle ear.
stimulus (stim′u-lus) agent that causes a change in the activity of a structure.
stress (stres) according to Selye, physiological stress is a condition in the body produced by all kinds of injurious factors that he calls "stressors" and manifested by a syndrome (a group of symptoms that occur together).
stressor (stres′sor) any injurious factor that produces biological stress; for example, emotional trauma, infections, severe exercise.
striated (stri′āt-id) marked with parallel lines.
sudoriferous (su″dor-if′er-us) secreting sweat.
sulcus (sul′kus) furrow or groove.
superior (su-pe′re-or) higher; opposite of inferior.
supinate (su′pĭ-nāt) to turn the palm of the hand upward; opposite of pronate.
Sylvius (sil′ve-us) seventeenth century anatomist.
symphysis (sim′fĭ-sis) Greek for growing together.
synapse (sin′aps) joining; point of contact between adjacent neurons.

synovia (si-no've-ah) literally "with egg"; secretion of the synovial membrane; resembles egg white.
synthesis (sin'thĭ-sis) putting together of parts to form a more complex whole.
systole (sis'to-le) contraction of the heart muscle.

talus (ta'lus) ankle; one of the bones of the ankle.
tarsus (tahr'sus) instep.
tendon (ten'dun) band or cord of fibrous connective tissue that attaches a muscle to a bone or other structure.
thorax (tho'raks) chest.
thrombosis (throm-bo'sis) formation of a clot in a blood vessel.
tibia (tib'e-ah) Latin for shin bone.
tonus (to'nus) continued, partial contraction of muscle.
tract (trakt) bundle of axons located within the central nervous system.
trauma (traw'mah) injury.
trochlear (trok'li-ar) pertaining to a pulley.
trophic (trof'ik) having to do with nutrition.
turbinate (tur'bĭ-nāt) shaped like a cone or like a scroll or spiral.
tympanum (tim'pah-num) drum.

umbilicus (um"bĭ-li'kus) navel.
utricle (u'tre-k'l) little sac.
uvula (u'vu-lah) Latin for a little grape; a projection hanging from the soft palate.

vagina (vah-ji'nah) sheath.
vagus (va'gus) Latin for wandering.
valve (valv) structure that permits flow of a fluid in one direction only.
vas (vas) vessel or duct.
vastus (vas'tus) wide; of great size.
vein (vān) vessel carrying blood toward the heart.
ventral (ven'tral) of or near the belly; in man, front or anterior; opposite of dorsal or posterior.
ventricle (ven'trĭ-k'l) small cavity.
vermiform (ver'mĭ-form) worm-shaped.
villus (vil'lus) hairlike projection.
viscera (vis'er-ah) internal organs.
vomer (vo'mer) ploughshare.

xiphoid (zi'foid) sword-shaped.

zygoma (zi-go'mah) yoke.

Index

C